A Level Biology for OCR A

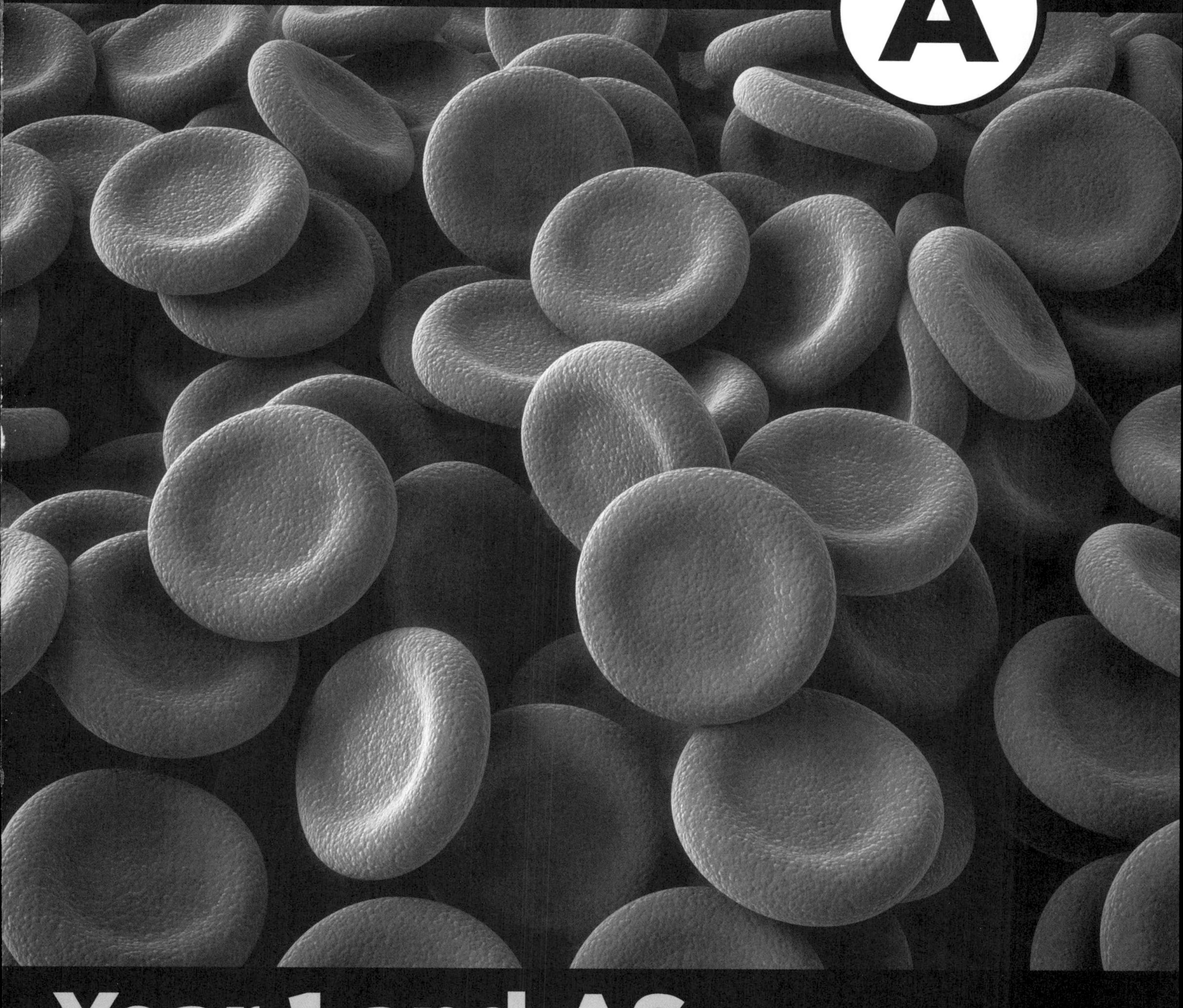

Year 1 and AS

Michael Fisher

OXFORD
UNIVERSITY PRESS

OXFORD
UNIVERSITY PRESS

Great Clarendon Street, Oxford, OX2 6DP, United Kingdom

Oxford University Press is a department of the University of Oxford. It furthers the University's objective of excellence in research, scholarship, and education by publishing worldwide. Oxford is a registered trade mark of Oxford University Press in the UK and in certain other countries

First published in 2016

British Library Cataloguing in Publication Data
Data available

978 0 19 835193 1

10 9 8 7 6 5 4 3 2 1

Paper used in the production of this book is a natural, recyclable product made from wood grown in sustainable forests. The manufacturing process conforms to the environmental regulations of the country of origin.

Printed in Great Britain by Bell and Bain Ltd, Glasgow.

Artwork by Q2A Media

FSC www.fsc.org MIX Paper from responsible sources FSC® C007785

AS/A Level course structure

This book has been written to support students studying for OCR AS Biology A and for students in their first year of studying for OCR A Level Biology A. It covers the AS modules from the specification, the content of which will also be examined at A Level. The modules covered are shown in the contents list, which also shows you the page numbers for the main topics within each module. If you are studying for OCR AS Biology A, you will only need to know the content in the shaded box.

AS exam

Year 1 content

1 Development of practical skills in biology
2 Foundations in biology
3 Exchange and transport
4 Biodiversity, evolution, and disease

Year 2 content

5 Communication, homeostasis, and energy
6 Genetics, evolution, and ecosystems

A level exam

A Level exams will cover content from Year 1 and Year 2 and will be at a higher demand.

How to use this book

This book contains many different features. Each feature is designed to support and develop the skills you will need for your examinations, as well as foster and stimulate your interest in biology.

Worked example

Step-by-step worked solutions.

Common misconception

Common student misunderstandings clarified.

Go Further

Familiar concepts in an unfamiliar context.

Maths skill

A focus on maths skills.

Questions and model answer

Sample answers to exam style-style questions.

Practical skill

Support for the practical knowledge requirements of the exam.

Chapter 2 Practice questions

1 The actual diameter of an *Amoeba proteus* cell is 0.25 mm. The cell is studied under a light microscope using a magnification of ×1600.

Which of the following values represents the image radius of the cell as seen under the microscope?

A 0.4 m

B 400 cm

C 200 mm

D 20 cm *(1 mark)*

2 Which of the following statements is / are true of the procedure you would use to differentiate between Gram-positive and Gram-negative bacteria?

1 A primary stain is applied to heat-fixed bacteria.

2 Ethanol can be added to decolourise Gram-negative bacteria.

3 A counterstain such as safranin is applied.

A 1, 2, and 3 are correct

B Only 1 and 2 are correct

C Only 2 and 3 are correct

D Only 1 is correct *(1 mark)*

3 Which of the following statements is true of a transmission electron microscope?

A A maximum resolution of 0.5 nm can be achieved.

B A maximum magnification of ×100 000 can be achieved.

C 3D images can be produced.

D The preparation of samples is non-invasive. *(1 mark)*

4 Complete the following table by writing YES or NO in the blank boxes.

Structure	Membrane-bound	Contains DNA	
Mitochondrion			*(1 mark)*
Chloroplast			*(1 mark)*
Lysosome			*(1 mark)*
Microtubules			*(1 mark)*

Practice questions at the end of each chapter including questions that cover practical and maths skills.

Specification references

At the beginning of each topic, there are specification references to allow you to monitor your progress.

Key term

Pulls out key terms for quick reference.

Revision tip

Prompts to help you with your understanding and revision.

Synoptic link

These highlight the key areas where topics relate to each other. As you go through your course, knowing how to link different areas of physics together becomes increasingly important. Many exam questions, particularly at A Level, will require you to bring together your knowledge from different areas.

Summary Questions

1 These are short questions at the end of each topic.

2 They test your understanding of the topic and allow you to apply the knowledge and skills you have acquired.

3 The questions are ramped in order of difficulty.

2.1 Microscopy

Specification reference: 2.1.1(a), (b), (c), and (d)

This chapter is concerned with the appearance of cells. Observing biological material is a fundamental scientific skill. The invention of the light microscope more than 400 years ago allowed cells to be observed. Since then, microscope technology has advanced to enable fine details to be seen. Here you will learn how a modern light microscope is used.

How does a light microscope work?

Light shines up though the sample being observed and through two lenses (**objective** and **eyepiece**). Each lens magnifies the image; the combined effect of the two lenses can increase an image by up to 2000 times its actual size.

Revision tip: The condenser lens

Light first passes through a condenser lens. This does not magnify; it focuses light on the sample being observed. You are not required to learn about condenser lenses.

Practical skill: Preparation of a sample

The type of biological material being studied will dictate how it is prepared for a light microscope. Specimens may need to be **sectioned** (cut into thin slices), **fixed** (preserved and sterilised), and **stained** (see below). Cover slips can be placed over specimens immersed in liquid (**wet mounting**) or without liquid (**dry mounting**).

Staining improves the visibility of structures within a specimen. Stains increase the contrast between different structures (because components take up stains in different ways). **Differential staining** can therefore distinguish between structures in cells or between different species (if their cells contain different components). For example, some bacteria retain crystal violet and appear purple (*Gram-positive bacteria*), whereas others do not (*Gram-negative bacteria*).

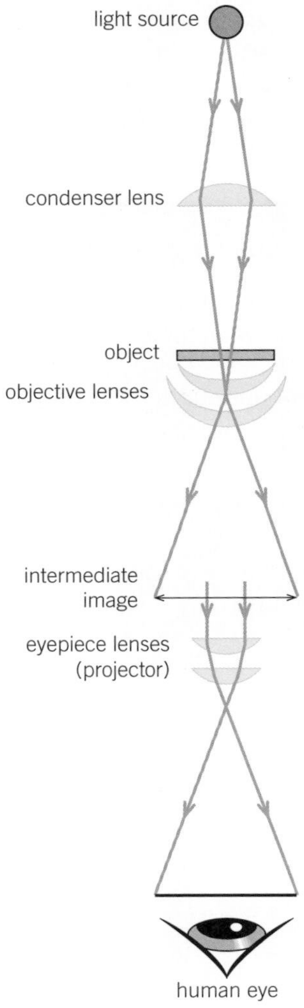

▲ **Figure 1** *The path of light through a microscope*

Revision tip: Drawing conclusions

Scientific drawings should be clear and precise; they should be produced using a 'less-is-more' principle (e.g. drawing smooth continuous lines, without unnecessary detail, and showing essential features).

Summary questions

1. Explain why a specimen needs to be sectioned before it is examined under a light microscope. *(1 mark)*
2. Describe the roles of the eyepiece and objective lenses in a light microscope. *(2 marks)*
3. FABIL stains lignin yellow and cellulose blue. Using knowledge from Topic 2.5, The ultrastructure of plant cells, and Topic 9.1, Transport systems in dicotyledonous plants, suggest the benefit of FABIL to a person observing plant tissue. *(3 marks)*

2.2 Magnification and calibration

Specification reference: 2.1.1(b), (e), and (f)

When observing biological material under a microscope, scientists need to know how much larger the image is in comparison to the actual size of the material. This is the concept of magnification, which we discuss here.

Magnification and resolution

Magnification is the extent to which the actual size of an object is enlarged into the image seen through a microscope. **Resolution** is the extent to which two objects can be distinguished as separate structures. More detail can be seen at higher resolutions.

Key term

Resolution: The shortest distance between two objects that are still seen as separate structures.

Revision tip: It can all be a blur

Magnification may increase without resolution improving. This will result in a larger image, but one that is blurred.

Maths skill: Calculating magnification

$$\text{Magnification} = \frac{\text{size of image}}{\text{actual size of object}}$$

Worked example: Rearranging the formula

You may be asked to calculate image size or the actual size of an object, rather than magnification. This requires the formula to be rearranged. For example, if a microscope has a magnification of ×1500, and a red blood cell appears to have a diameter of 1.05 cm, how do we calculate the actual diameter of the cell?

$$\text{Actual size} = \frac{\text{size of image}}{\text{magnification}} = \frac{1.05}{1500} = 0.0007\ \text{cm}\ (7\ \mu\text{m})$$

Revision tip: How significant?

Check the number of significant figures required in an answer. A calculator answer of 5.478 μm would be 5.5 μm to two significant figures, for example.

Revision tip: If you forget the formula …

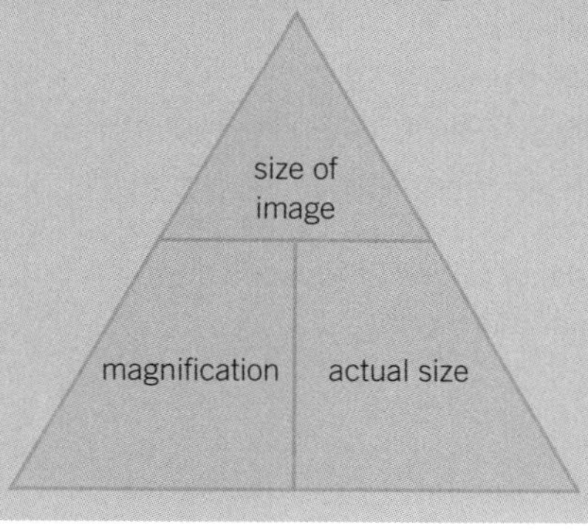

Mnemonics are useful tools for remembering the arrangement of formula triangles. For example, you can use 'I'm Mag actually' to remember the arrangement of the magnification formula.

Practical skill: Calibrating microscopes

A microscope eyepiece contains a scale without divisions (i.e. a graticule). Calibrating a microscope involves working out the length represented by each division in this scale at a particular magnification. This enables you to measure the sample being observed.

The graticule scale is compared to a stage micrometer (a slide with a scale in μm). You can then calculate the number of μm per eyepiece division. For example, each division on the Figure 1 eyepiece scale represents 30 μm (12 stage micrometer units = 120 μm, divided by four eyepiece units = 30 μm).

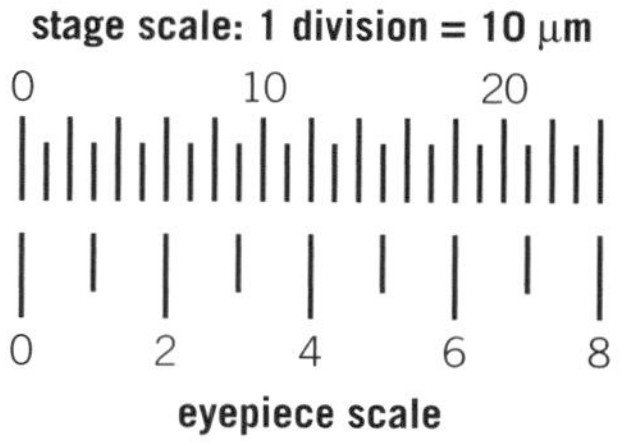

▲ **Figure 1** *Graticule and micrometer scales are compared*

Summary questions

1 State the difference between magnification and resolution. *(2 marks)*

2 Calculate the diameter of a bacterium that appears 2.4 mm wide under a microscope (×1600 magnification). *(2 marks)*

3 An erythrocyte has a diameter of 7 μm. Calculate the image size when viewed at magnification ×1600 under a light microscope. Express your answer in standard form and to two significant figures. *(3 marks)*

2.3 More microscopy

Specification reference: 2.1.1(a) and (f)

The study of cell details was made possible by the invention of the light microscope. Nowadays, scientists have access to microscopes that are much more powerful. Rather than using regular visible light, these microscopes use electron beams and lasers, allowing a wealth of detail to be visualised within cells.

Comparing microscopes

Electron beams have shorter wavelengths than light waves. This is why electron microscopes have greater resolutions than light microscopes. Electron microscopes work by either transmitting electrons through a sample (TEM) or scanning the sample (SEM). The two microscopes have different properties, as shown in the following table.

	Light	Electron microscope	
		Transmission (TEM)	Scanning (SEM)
How does it work?	See Topic 2.1, Microscopy	Electron beams pass through the specimen	Electrons are reflected back from the specimen and detected
Magnification	Up to ×2000	Up to ×500 000	Up to ×100 000
Resolution (nm)	200	0.5	3–10
Benefits	Inexpensive, with short preparation time Colours and living material can be observed	High magnification and resolution Enables intracellular details to be observed	High magnification and resolution 3D images
Disadvantages	Limited magnification and resolution	Expensive, with complex preparation of samples. Images are black and white (but artificial colour can be added). Artefacts can be introduced during preparation.	

Revision tip: Artefacts

An artefact is something visible in a microscope image that is not a natural part of the specimen (e.g. structural distortion created during preparation).

Synoptic link

See Topic 2.1, Microscopy, for a reminder of the approach to scientific drawing.

A modern light microscope

The confocal laser scanning microscope (CLSM) uses laser light to scan a specimen. Light is absorbed by fluorescent chemicals and radiated back from the specimen. A laser is a device that produces light in which the waves are lined up together in the same direction. As a result, light from lasers is very bright and can be focused on a small spot.

Although the resolution of these microscopes is poor compared to electron microscopes, they have two benefits:

- *3D images* are produced
- They are *non-invasive*, which allows living tissue to be observed.

Practical skill: Drawing from electron micrographs

Electron microscopes show more detail than light microscopes because of their greater resolution. The principles you should follow when drawing images are the same in either context.

Summary questions

1 Describe and explain the difference in the resolving power of electron and light microscopes. *(2 marks)*

2 A chloroplast with a diameter of 6 μm was magnified using a transmission electron microscope. The magnification was set to ×50 000. Calculate the size of the image. Express your answer in standard form. *(2 marks)*

3 Evaluate the advantages and disadvantages of electron microscopes and CLSMs. *(5 marks)*

2.4 Eukaryotic cell structure

Specification reference: 2.1.1(g), (i), and (j)

Thanks to microscopy, scientists now have an excellent understanding of the contents and structure of cells. The architecture of cells, however, can differ between species. You will examine some of these differences in the following three topics. We begin by discussing eukaryotic cells.

Cellular components of eukaryotic cells

Species consist of either prokaryotic or eukaryotic cells (see Topic 2.6, Prokaryotic and eukaryotic cells). Animals, fungi, plants, and protoctista are eukaryotes. The most significant feature of eukaryotic cells is that they contain compartments called **organelles**.

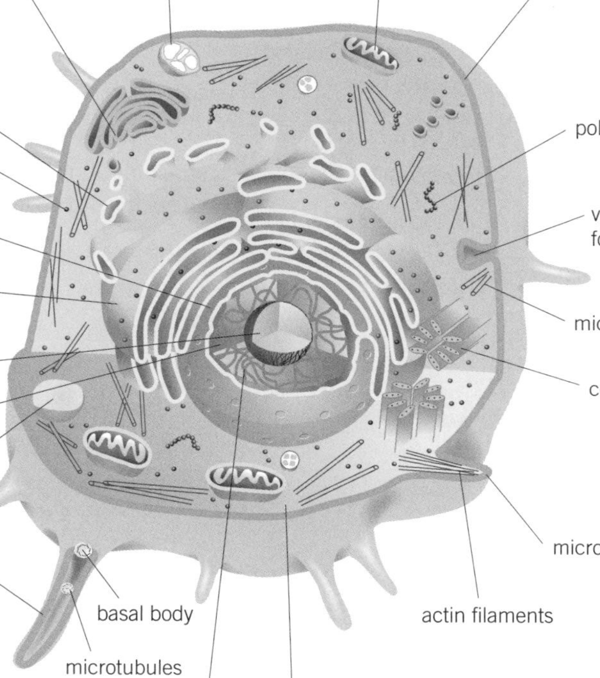

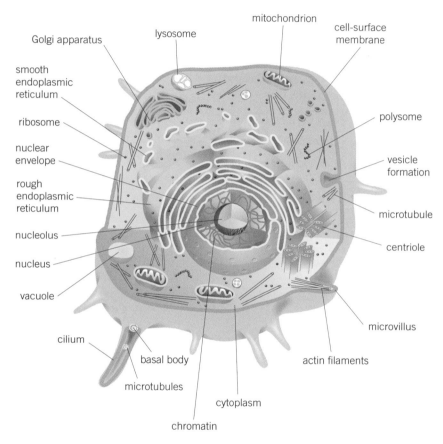

▲ **Figure 1** *The generalised ultrastructure of an animal eukaryotic cell*

The following table outlines the functions and features of membrane-bound organelles and cell components lacking membranes in eukaryotic cells.

Component	Function	Key features
Nucleus (plural: nuclei)	Contains genetic information	Surrounded by a **nuclear envelope** (containing pores) Contains **chromatin** (DNA + histone proteins), which condenses to form chromosomes Contains a **nucleolus** (which produces rRNA)
Mitochondria (singular: mitochondrion)	ATP production through aerobic **respiration**	Double membrane; the inner membrane folds to form **cristae** Internal fluid is called the **matrix**

Synoptic link

You will learn more about the classification and naming of organisms in Topic 10.2, The five kingdoms.

Revision tip: Ultrastructure

The ultrastructure of a cell is the fine detail that can be observed only by using an electron microscope.

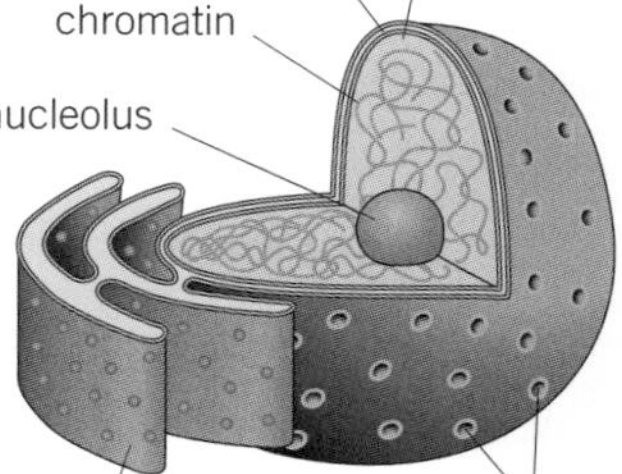

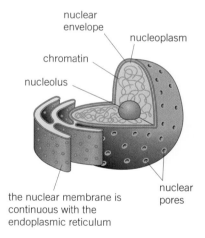

▲ **Figure 2** *The relationship between the nucleus and endoplasmic reticulum*

Key term

Organelles: Structures in eukaryotic cells with specific roles. (They are often defined as being membrane-bound, although some scientists include structures without membranes.)

Revision tip: Cytosol vs cytoplasm

Cytosol is the fluid within a cell. Cytoplasm comprises the cytosol plus the organelles suspended within it. The one exception is the nucleus, which is considered to be separate to the cytoplasm.

Synoptic link

Chapter 3, Biological molecules, delves further into DNA structure and protein synthesis.

Lysosomes	Breaking down **waste** (e.g. old organelles)	Specialised **vesicles** (membranous sacs), containing **hydrolytic enzymes**
Smooth endoplasmic reticulum (ER)	**Lipid** and carbohydrate synthesis	Flattened membrane-bound sacs (**cisternae**)
Rough endoplasmic reticulum (ER)	**Protein synthesis**	Cisternae bound to **ribosomes** (which are made of RNA; ribosomes can also appear loose in cytoplasm)
Golgi apparatus	**Modifying proteins** and packaging them into vesicles	Flattened sacs (similar to smooth ER)
Cytoskeleton	Maintaining cell shape Control of cell movement and of organelle movement in cells Compartmentalisation of organelles	**Microfilaments** (made from actin) control cell movement and cytokinesis (see Topic 6.2, Mitosis). **Microtubules** (made from tubulin) regulate shape and organelle movements. They form centrioles and spindle fibres (see Topic 6.2, Mitosis).
Flagella (singular: flagellum) and cilia (singular: cilium)	Flagella enable cell movement. Cilia move substances across cell surfaces.	Both comprise a cylinder containing 11 **microtubules** (nine in a circle and two in the centre). Flagella are longer than cilia.
Cell surface (plasma) membrane	See Topic 5.1, The structure and function of membranes	

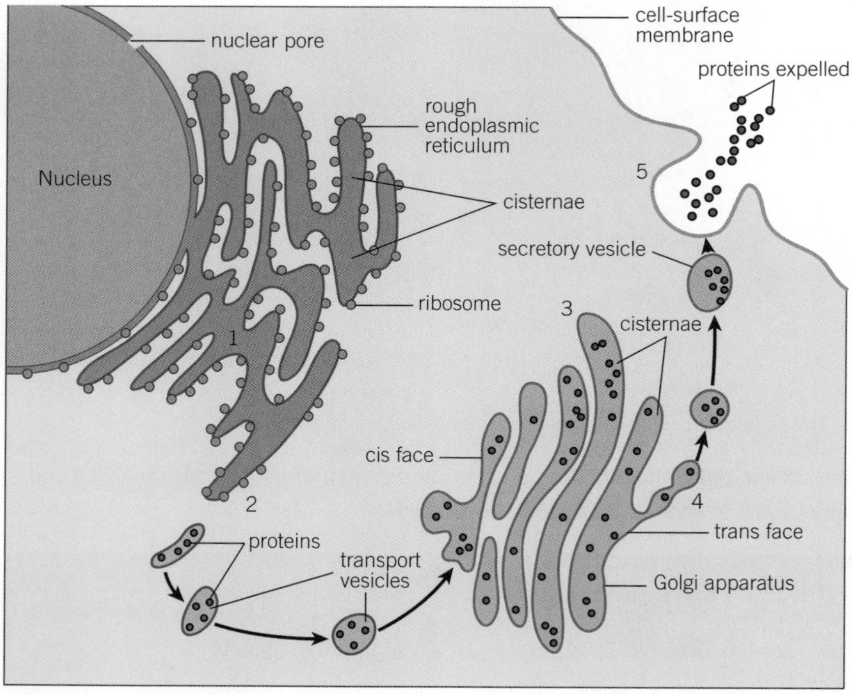

▲ **Figure 3** *Protein production and processing: (1) Synthesis in rough ER. (2) Transport in vesicles. (3) Structural modification in the Golgi apparatus. (4) Modified proteins leave the Golgi in vesicles. (5) Some proteins are released from the cell (others are retained).*

Summary questions

1 State one structural similarity and one structural difference between a flagellum and a cilium. *(2 marks)*

2 Describe the stages undergone by a protein from its production to its release from a cell. *(5 marks)*

3 Actin microfilaments can form networks that assemble and disassemble rapidly within cells. Suggest how this activity benefits some cells. *(3 marks)*

2.5 The ultrastructure of plant cells

Specification reference: 2.1.1(g)

Plant cell features

Plant cells are eukaryotic and tend to possess the components we discussed in Topic 2.4, Eukaryotic cell structure (although centrioles are absent from flowering plants). Plants, in addition, have three structures that other eukaryotic cells lack: cellulose cell walls, large vacuoles, and chloroplasts.

Component	Description	Function
Cellulose cell wall	A strong barrier on the outside of the cell membranes. It is composed of a polysaccharide called **cellulose**.	Provides shape and rigidity
Large, permanent vacuole	A large, central sac (surrounded by a membrane known as a **tonoplast**). It principally contains water.	Increases **turgor** (i.e. the extent to which cells are filled with water)
Chloroplasts	An organelle containing fluid (**stroma**) and a network of membranes (**thylakoids**, which are stacked as **grana**).	The organelle responsible for **photosynthesis**

Synoptic link

You will learn about the structure of cellulose in Topic 3.3, Carbohydrates.

Revision tips: Walls and vacuoles in other organisms

Cell walls are found in organisms other than plants. Fungi possess complex walls containing chitin. Bacteria have walls made of peptidoglycan.

All fungal cells have at least one vacuole, although they display some functional differences to plant vacuoles. Some bacterial cells contain vacuoles.

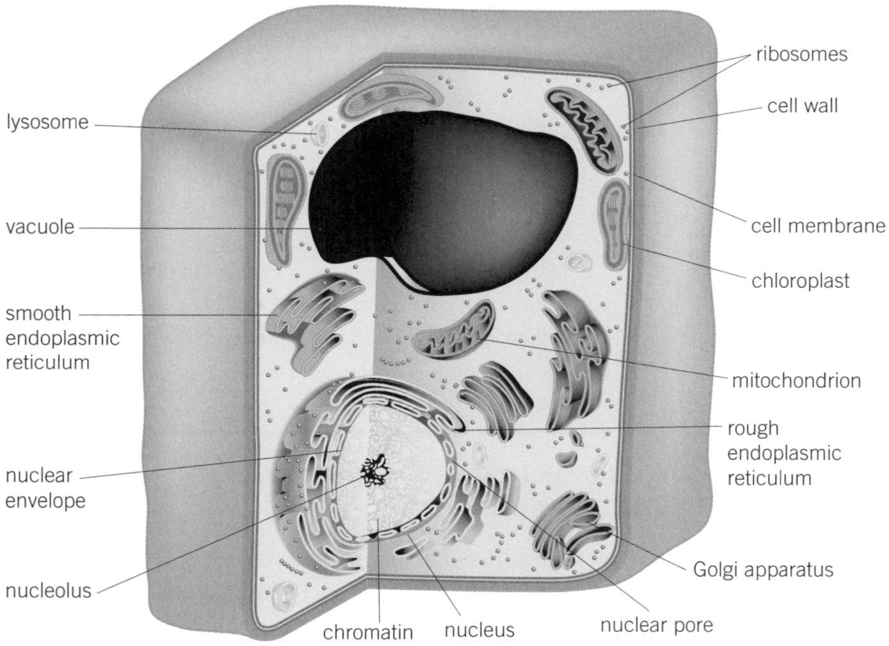

▲ **Figure 1** *A generalised plant cell*

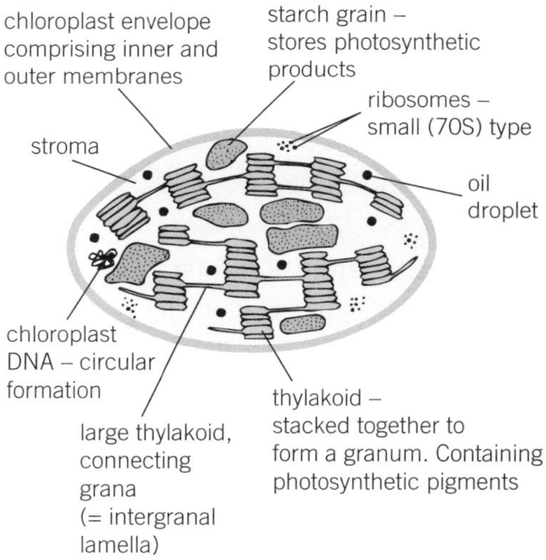

▲ **Figure 2** *Chloroplast structure*

Revision tips: Standard advice

The ability to convert measurements between decimal and standard form is an important skill. The following examples provide a good comparison.

Chloroplast length

$8\,\mu m = 0.000008\,m = 8 \times 10^{-6}\,m$ (standard form)

Palisade cell diameter

$40\,\mu m = 0.000040\,m = 4 \times 10^{-5}\,m$ (standard form)

Summary questions

1 State the principal role of a vacuole in plant cells. *(1 mark)*

2 *Chloroplasts are the only truly unique feature of plant cells.* Evaluate the validity of this statement. *(2 marks)*

3 Ribosomes in chloroplasts are similar to the 70S ribosomes of prokaryotic cells, but with some structural differences. What does this suggest about the possible evolutionary history of chloroplasts? *(3 marks)*

2.6 Prokaryotic and eukaryotic cells

Specification reference: 2.1.1(k)

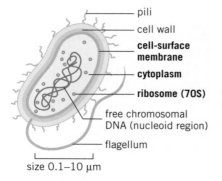

▲ **Figure 1** *A generalised prokaryotic cell; the features in bold are found in all eukaryotic cells as well*

Revision tip: S for … size?

S is a unit that indicates how quickly a substance settles at the bottom of a centrifuge tube (centrifugation is a method for separating cell contents). In terms of ribosomes, we can think of *S* as a size indicator (i.e. 70S ribosomes are smaller than 80S).

Summary questions

1 State the difference in the structure of prokaryotic and eukaryotic cell walls. *(2 marks)*

2 Bacteria possess pili, which are hair-like structures that allow the exchange of genetic material between cells. Suggest why this is useful for bacteria. *(2 marks)*

3 Describe the differences in DNA packaging between prokaryotic and eukaryotic cells. *(4 marks)*

You learned about the ultrastructure of eukaryotic cells in Topic 2.4, Eukaryotic cell structure, and Topic 2.5, The ultrastructure of plant cells. Here you will learn about prokaryotic cells and examine the differences between the two categories of cell.

How are prokaryotic cells different?

Prokaryotes are always unicellular, whereas many eukaryotic species are composed of more than one cell.

These two types of cell also exhibit many structural differences, as shown in the following table.

Feature	Prokaryotic cells	Eukaryotic cells
Membrane-bound organelles	No	Yes
DNA	One main molecule of DNA (**circular** and found naked in the cytoplasm) Additional DNA can be found in **plasmids**	Linear (**chromosomes**) and associated with histone proteins in a nucleus Additional DNA found in mitochondria and chloroplasts
Cytoskeleton	Yes (but comprising different proteins to the eukaryotic cytoskeleton)	Yes
Ribosomes	Smaller (70S)	Larger (80S)
Cilia and flagella	Flagella are sometimes present, but with a different structure to eukaryotes	Often present in animal cells
Cell wall	Yes, made of peptidoglycan (murein), a polymer of sugars and amino acids	Present in plant cells (made of cellulose) and fungi (made of chitin)
Reproduction	Asexual (binary fission)	Asexual or sexual

Go further: How do prokaryotes cope without membrane-bound organelles?

You might be wondering how bacteria carry out aerobic respiration without mitochondria. Approximately 50 years ago, infoldings of bacterial plasma membranes were observed under electron microscopes. These were called mesosomes and proposed as the prokaryotic equivalent of mitochondrial cristae. They are now thought to be artefacts (not occurring naturally in bacterial cells).

1 Suggest how the infoldings of bacterial plasma membranes may have been produced.

2 Suggest how aerobic respiration can occur in bacteria.

Chapter 2 Practice questions

1 The actual diameter of an *Amoeba proteus* cell is 0.25 mm. The cell is studied under a light microscope using a magnification of ×1600.

Which of the following values represents the image radius of the cell as seen under the microscope?

A 0.4 m

B 400 cm

C 200 mm

D 20 cm *(1 mark)*

2 Which of the following statements is / are true of the procedure you would use to differentiate between Gram-positive and Gram-negative bacteria?

1 A primary stain is applied to heat-fixed bacteria.

2 Ethanol can be added to decolourise Gram-negative bacteria.

3 A counterstain such as safranin is applied.

A 1, 2, and 3 are correct

B Only 1 and 2 are correct

C Only 2 and 3 are correct

D Only 1 is correct *(1 mark)*

3 Which of the following statements is true of a transmission electron microscope?

A A maximum resolution of 0.5 nm can be achieved.

B A maximum magnification of ×100 000 can be achieved.

C 3D images can be produced.

D The preparation of samples is non-invasive. *(1 mark)*

4 Complete the following table by writing YES or NO in the blank boxes.

Structure	Membrane-bound	Contains DNA	
Mitochondrion			*(1 mark)*
Chloroplast			*(1 mark)*
Lysosome			*(1 mark)*
Microtubules			*(1 mark)*

5 The following passage describes the preparation of samples for examination under a light microscope. Complete the passage by choosing the most appropriate word to place in each gap.

Samples can be preserved by .... with chemicals such as formaldehyde. .... staining enables tissues and organelles within a sample to be distinguished. For example, Gram staining involves the application of .... to identify bacteria based on the structure of their .... *(4 marks)*

6 The cell body of a motor neuron had a diameter of three eyepiece units. What is the diameter of the cell body in standard form? *(2 marks)*

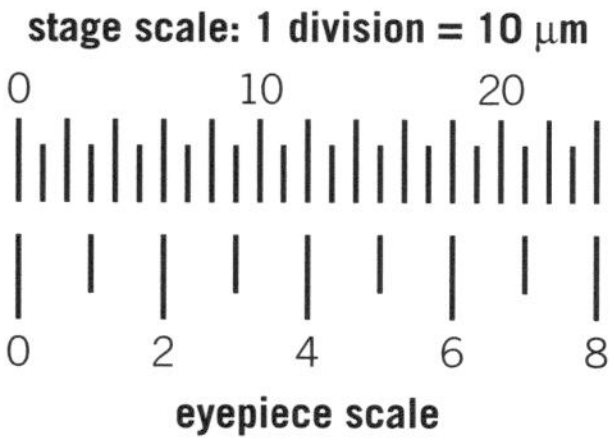

7 Compare the structure and function of flagella in eukaryotic and prokaryotic cells. *(3 marks)*

3.1 Biological elements

Specification reference: 2.1.2(b), (c), and (p)

Revision tip: The long and the short of it

Polymers are long chains of smaller molecules (**monomers**) that have been bonded together. We will discuss examples of monomers and polymers in subsequent topics.

Revision tip: Organic learning

Organic substances contain carbon bonded to hydrogen; inorganic substances lack C–H bonds. Hydrogen carbonate (HCO_3^-) is considered inorganic because its carbon atom is bonded only to oxygen.

In this chapter you will learn about a range of important biological molecules. First you will look at the chemistry of the molecules you will be studying. You will also learn about some of the common ions found in biological systems.

The elements in biological molecules

The following table shows the elements that are always present in biological molecules or sometimes present.

Element	Carbon (C)	Hydrogen (H)	Oxygen (O)	Nitrogen (N)	Phosphorus (P)	Sulfur (S)
Atomic (proton) number	6	1	8	7	15	16
Typical number of bonds formed	4	1	2	3	3–5	2
Carbohydrates (Topic 3.3)	always present	always present	always present			
Lipids (Topic 3.5)	always present	always present	always present			
Proteins (Topic 3.6)	always present	always present	always present	always present		sometimes present
Nucleic acids (Topic 3.8)	always present	always present	always present	always present	always present	

Common misconception: Molecules and compounds

An element is a substance composed of one type of atom (e.g. pure carbon consists only of atoms with six protons and six electrons). Molecules contain more than one atom and can be either elements (e.g. O_2 or N_2) or compounds (which comprise more than one element, for example glucose – $C_6H_{12}O_6$). Compounds contain either ionic bonds between elements (e.g. NaCl) or covalent bonds (e.g. $C_6H_{12}O_6$). Ionic compounds are *not* molecules. They have giant lattice structures.

Inorganic ions

Ions are atoms or molecules that have lost electrons (positively charged ions – cations) or gained electrons (negatively charged ions – anions). Several inorganic ions play essential roles in organisms.

▼ **Table 1** *Inorganic ions in biology*

Ion	Examples of roles	Topic reference
Ca^{2+} (calcium)	Nervous impulse transmission	Action potentials (Year 2)
Na^+ (sodium)		
K^+ (potassium)		
H^+ (hydrogen)	Determination of pH in solutions	4.2, Factors affecting enzyme activity
NH_4^+ (ammonium)	Sources of nitrogen for plants	The nitrogen cycle (Year 2)
NO^{3-} (nitrate)		
HCO_3^- (hydrogen carbonate)	Transport of respiratory gases	8.4, Transport of oxygen and carbon dioxide in the blood
Cl^- (chloride)		
PO_4^{3-} (phosphate)	Nucleic acid and ATP formation	3.8, Nucleic acids, and 3.11, ATP
OH^- (hydroxide)	Determination of pH in solutions	4.2, Factors affecting enzyme activity

Summary questions

1. State the three elements found in carbohydrates, proteins, lipids, and nucleic acids. *(3 marks)*
2. Explain the difference between a compound and a molecule. *(3 marks)*
3. State which of the ions in the list below are **a** inorganic ions **b** organic ions **c** compound ions Cl^-, HCO_3^-, NH_4^+, CH_3COO^-, Ca^{2+}, OH^- *(3 marks)*

3.2 Water

Specification reference: 2.1.2(a)

Molecules interact with each other in a variety of ways, depending on the atoms they contain. Water molecules form hydrogen bonds with each other, which gives water the properties that make it crucial for life.

Hydrogen bonding

Water molecules (H_2O) are polar (i.e. the oxygen atom has a partial negative charge and the hydrogen atoms have partial positive charges). As a consequence, the oxygen atom in a water molecule is attracted to hydrogen atoms in neighbouring molecules. This attraction is called a **hydrogen bond**.

Properties of water

Hydrogen bonds are strong in comparison to other intermolecular forces. The unique properties of water are a result of hydrogen bonding. These properties enable water to play a pivotal role in biology.

▼ **Table 1** *The properties of water*

Property	Explanation	Importance
Good solvent	Polar water molecules attract (and dissolve) other polar molecules and ions	Water transports dissolved solutes (e.g. in blood and phloem) Chemical reactions occur in water
High specific heat capacity	A relatively large amount of energy is required to increase water temperature	Thermal stability in aquatic environments and inside organisms
High heat of vaporisation	Additional energy is needed to change water from liquid to gas	Thermoregulation – sweating and panting can cool an organism when water on the body is evaporated
Cohesion	Hydrogen bonds cause water molecules to be attracted to each other and flow together	Water movement up xylem vessels Surface tension – small organisms can move on the water surface
Low density solid (i.e. ice)	The crystalline structure in ice is less dense than liquid water	Provides an insulating layer for aquatic habitats in cold climates The ice surface provides a habitat for some organisms (e.g. polar bears)

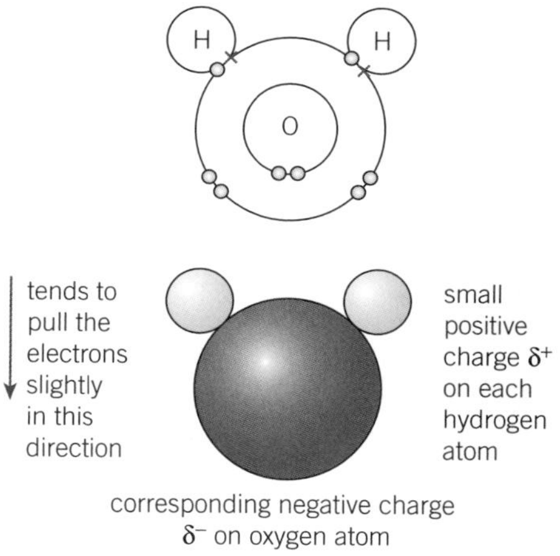

▲ **Figure 1** *Water is a polar molecule*

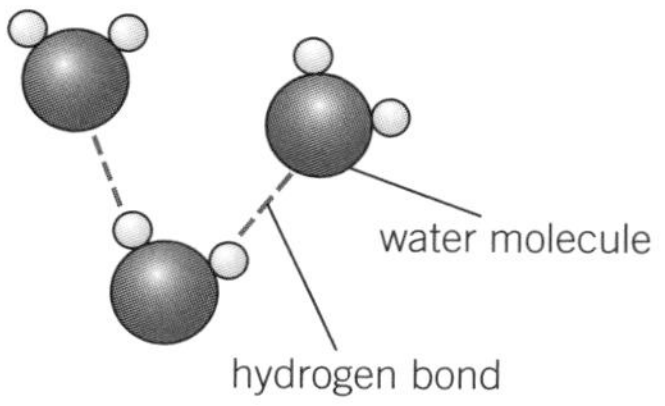

▲ **Figure 2** *Hydrogen bonds between water molecules*

Synoptic link

You will study water cohesion and adhesion in xylem vessels in Topic 9.3, Transpiration, and its role as a solvent in Topic 8.3, Blood, tissue fluid, and lymph.

Summary questions

1 Explain why water is a polar molecule. *(3 marks)*

2 Explain how the properties of water are suited to its role as a transport medium in blood and xylem vessels. *(3 marks)*

3 Amino acids exist as zwitterions in solution (i.e. each amino acid molecule has both positive and negative charges). Explain why zwitterions dissolve in water. *(2 marks)*

3.3 Carbohydrates

Specification reference: 2.1.2(d), (e), (f), and (g)

> **Revision tip: Alpha or beta?**
> α-glucose and β-glucose are known as isomers. They have the same molecular formula ($C_6H_{12}O_6$) but a slightly different arrangement of atoms. To be specific, the only difference between the two molecules is the position of H and OH on carbon 1.

Carbohydrates are molecules containing only carbon, oxygen, and hydrogen atoms. The chemical energy stored in the bonds of some carbohydrates can be used by organisms. Other carbohydrates have structural and storage roles. Here you will learn about a range of carbohydrates vital to biology.

Monosaccharides

A monosaccharide is the simplest carbohydrate unit (monomer). Two examples of monosaccharides that you will encounter in future topics are **glucose** and **ribose**.

▼ **Table 1** *The structures and functions of glucose and ribose*

Monosaccharide	Number of carbon atoms	Formula	Structure	Use
Glucose	6 (hexose sugar)	$C_6H_{12}O_6$	α-glucose 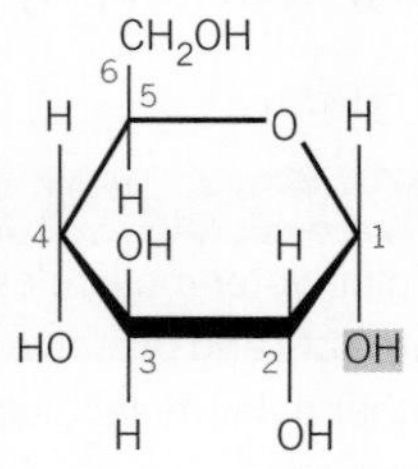 β-glucose 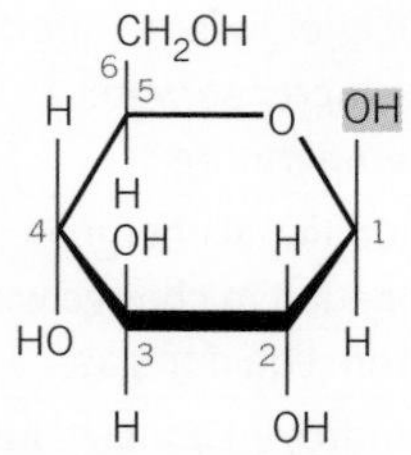	**α-glucose:** a substrate in respiration (a Year 2 topic) **β-glucose:** polymerises to form cellulose (see 'Polysaccharides')
Ribose	5 (pentose sugar)	$C_5H_{10}O_5$	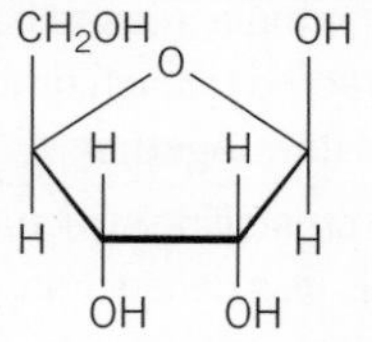	The sugar in RNA nucleotides (Topic 3.8, Nucleic acids) and ATP (Topic 3.11, ATP)

Disaccharides

Two monosaccharides can react together to form a disaccharide. This reaction is called a **condensation reaction**; it creates a **glycosidic bond** between the two monosaccharides and produces water. Disaccharides can be broken down to reform the original two monosaccharides in a **hydrolysis** reaction.

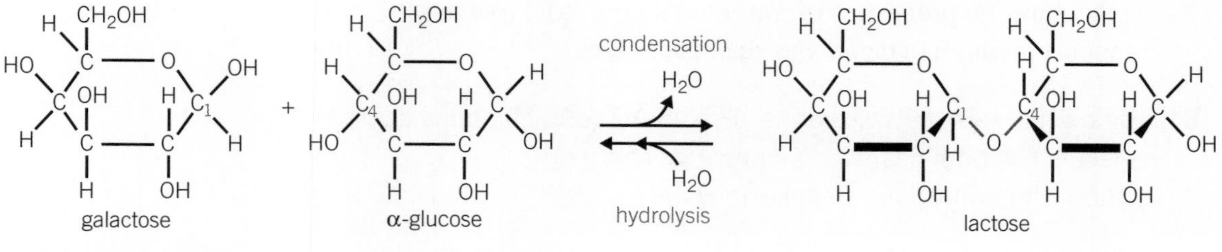

▲ **Figure 1** *The formation and breakdown of lactose*

▼ **Table 2** *The three disaccharide sugars and their constituent monomers*

Monosaccharides	Disaccharide produced
Glucose + glucose	Maltose
Glucose + galactose	Lactose
Glucose + fructose	Sucrose

Polysaccharides

Polysaccharides are long carbohydrate molecules (polymers) formed when many monosaccharides bond together in condensation reactions. Four examples of polysaccharides formed from glucose monosaccharides are outlined in Table 3.

▼ **Table 3** *The polysaccharides made from glucose monomers*

	Glycogen	Starch		Cellulose
		Amylose	Amylopectin	
Monomer	α-glucose	α-glucose		β-glucose
Type of glycosidic bonds	1,4 and 1,6 links	1,4 links	1,4 and 1,6 links	1,4 links
Branching	Yes	No	Yes	No
Helical	Yes	Yes	No	No
Function	Carbohydrate storage in animals	Carbohydrate storage in plants		Structural support in plant cell walls
Properties that suit function	Insoluble. Compact due to branching. The number of points at which glucose can be released through hydrolysis is increased by branching.	Amylose is insoluble. Helices and branching make starch compact. Branching in amylopectin increases the number of points at which glucose can be released through hydrolysis.		Insoluble. Cross links (hydrogen bonds between chains) increase strength.

Revision tip: Sugars, short and sweet

Polysaccharides are not classed as sugars, which are short-chain carbohydrates (i.e. monosaccharides and disaccharides).

Summary questions

1 Describe how sucrose is formed. *(3 marks)*

2 Describe two similarities and one difference between the structures of glucose and ribose. *(3 marks)*

3 Contrast the structures of cellulose and glycogen. Explain how the two structures are suited to the functions of the polysaccharides. *(6 marks)*

3.4 Testing for carbohydrates

Specification reference: 2.1.2(q) and (r)

You have already learned about a variety of carbohydrates. Now you will examine how to test for the presence of these carbohydrates.

Practical skill: Chemical tests

▼ **Table 1** *Chemical tests for starch, reducing sugars and non-reducing sugars*

	Benedict's test		Iodine test
Carbohydrate being identified	Reducing sugars (e.g. monosaccharides, lactose, maltose)	Non-reducing sugars (e.g. sucrose)	Starch
Description	Mix with Benedict's reagent in a boiling tube and heat	After a negative result with the Benedict's test, boil with dilute HCl. Conduct the Benedict's test a second time.	Mix iodine/potassium iodide solution with the sample
Negative result	**BLUE**		**YELLOW/BROWN**
Positive result (i.e. the carbohydrate is present)	**LOW = GREEN** **MEDIUM = ORANGE** **HIGH = RED**		**PURPLE/BLACK**

Revision tip: Reagent strips

An alternative to the Benedict's test and colorimetry is the use of reagent strips, which can be used to test for reducing sugars. The concentration of sugar is assessed by comparing the strip colour to a chart.

Colorimetry

A colorimeter can be used to assess the concentration of sugar in a solution.

Practical skill: Using a colorimeter following the Benedict's test

1. Insert a **red filter**.
2. Use a cuvette of distilled water to zero (**calibrate**) the colorimeter.
3. Construct a **calibration curve** (i.e. perform the Benedict's test on a series of solutions with known glucose concentrations, filter the precipitate, obtain a colorimeter reading (either transmission or absorption) for each solution, and plot a graph of transmission (or absorption) against concentration).
4. Conduct the Benedict's test on your test solution, then **filter the precipitate**.
5. Obtain a colorimeter reading for the solution and calculate the glucose concentration from the calibration curve.

Practical skill: Biosensors

Biosensors represent another method for determining the concentration of sugars in a solution. These machines require a recognition molecule that will bind the carbohydrate being assessed. The extent to which the carbohydrate binds determines the reading on the biosensor display. Unlike the Benedict's test, biosensors detect specific sugars (e.g. glucose).

Summary questions

1 State three quantitative methods for determining the concentration of glucose in a solution. *(3 marks)*

2 Sketch a colorimeter calibration curve for glucose concentration against **a** absorption **b** transmission. Assume that the precipitate is removed after the Benedict's test. Explain the shape of the absorption graph. *(6 marks)*

3 In colorimetry, a red filter is used when measuring transmission through solutions of Benedict's reagent. Explain why. *(2 marks)*

3.5 Lipids

Specification reference: : 2.1.2(h), (i), (j), and (q)

Like carbohydrates, lipids contain carbon, oxygen, and hydrogen. Unlike many carbohydrates, lipids are non-polar and largely insoluble in water, which means the two types of molecule occupy different roles in organisms.

Synoptic link

In Topic 5.1, The structure and function of membranes, you will learn more about the roles of phospholipids and cholesterol in membranes.

Examples of lipids

Lipids vary widely in structure. Three examples are triglycerides, phospholipids, and cholesterol.

▼ **Table 1** *Structures and functions of three lipids*

	Triglyceride	Phospholipid	Cholesterol
Structure	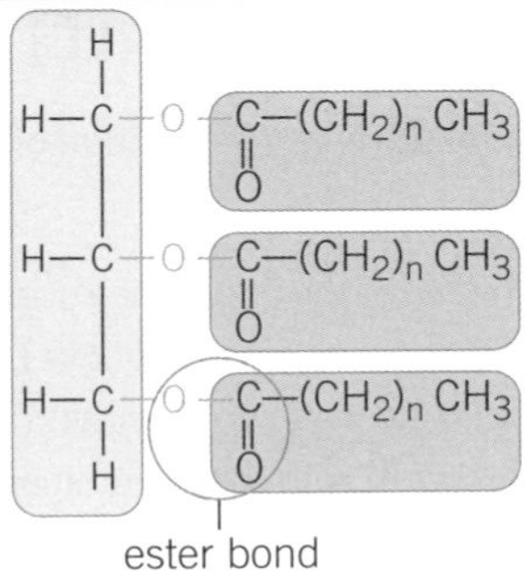	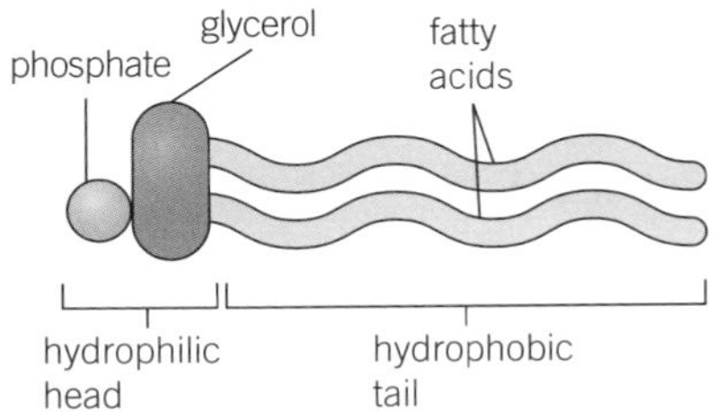	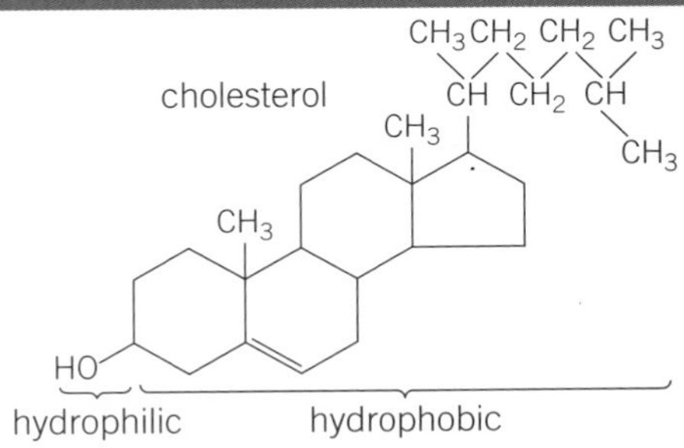
Properties	Compact and insoluble	Hydrophilic head and hydrophobic tail	Small; both hydrophilic and hydrophobic
Roles	Energy storage and insulation	Membrane structure	Membrane stability and steroid hormones

Synthesis of triglycerides

A triglyceride comprises a **glycerol** molecule and **three fatty acids**. Each fatty acid undergoes a **condensation reaction** with one of the OH (alcohol) groups in glycerol. A **hydrolysis** reaction breaks down the triglyceride into the original glycerol and fatty acids.

Revision tip: Saturated vs unsaturated

Fatty acids can be either saturated (no C=C double bonds) or unsaturated (at least one C=C double bond). Evidence exists that excessive consumption of saturated triglycerides raises the risk of coronary heart disease.

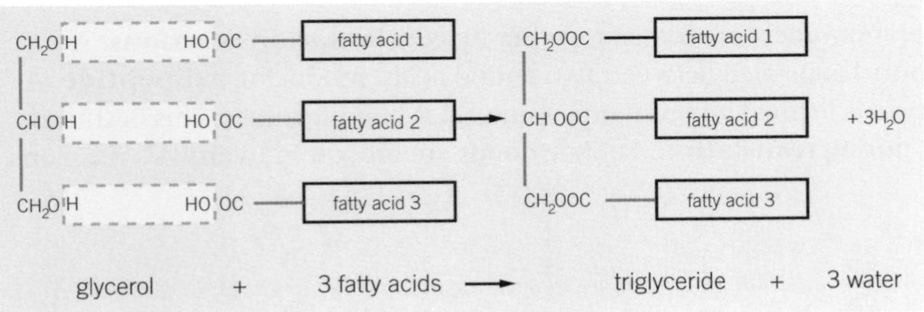

▲ **Figure 1** *Triglyceride synthesis via condensation reactions*

Summary questions

1 State what would be seen when an emulsion test gives a positive result. *(2 marks)*

2 Explain why the structure of a phospholipid is suited for its role in membranes. *(4 marks)*

3 Outline the similarities and differences between the synthesis of polypeptides and the synthesis of triglycerides. *(4 marks)*

Practical skill: The emulsion test

Lipids are identified using the emulsion test.

1. Mix your sample with **ethanol**
2. Mix with **water** and **shake**
3. The formation of a **white emulsion** on top of the solution indicates the presence of a lipid.

3.6 Structure of proteins

Specification reference: 2.1.2(k), (l), (m), (q), and (s)

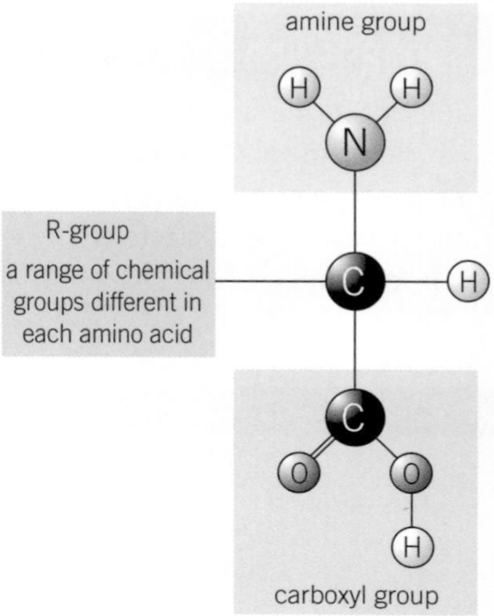

▲ **Figure 1** *The structure of an amino acid*

Synoptic link

You will learn more about polypeptide formation during translation in Topic 3.10, Protein synthesis.

Practical skill: The biuret test

Proteins are identified using the biuret test. This is a relatively simple test that does not require heat.

1. Add sodium hydroxide (NaOH) to the sample (equal volumes).
2. Add drops of copper sulphate solution (which is blue) and mix.
3. A purple colour shows the presence of protein.

Revision tip: Primary structure determines secondary structure

A protein's primary structure (the sequence of amino acids in a polypeptide) determines how the polypeptide folds or coils. The arrangement of R groups in the amino acid sequence dictates which bonds form and their locations. This influences both secondary and tertiary structure.

Proteins consist of amino acid monomers bonded together in chains. Here you will learn about the structure of amino acids. We also discuss how the secondary, tertiary, and quaternary structures of proteins produce their specific shapes.

Amino acids

All amino acids contain a central carbon atom, an amine group, a carboxyl group, and a hydrogen atom. The identity of an amino acid is determined by its R group; this is different in each of the 20 amino acids found in organisms.

Practical skill: Thin layer chromatography (TLC)

What does TLC do? It separates particles in a mixture (e.g. a solution of different amino acids).

How does TLC work? The mixture is applied at the base (the *origin*) of a silica gel layer (*stationary phase*), which is dipped into an organic solvent (*mobile phase*). Amino acids dissolve in the solvent as it diffuses up the gel. The rate at which the amino acids move up the gel depends on how much they interact (*adsorb*) with the gel. Amino acids will interact to different extents and will therefore move different distances.

How can the amino acids be visualised? The position of the amino acids can be seen by spraying ninhydrin onto the gel. This reveals each set of amino acids as a purple/brown spot.

How are the amino acids identified? The identity of each amino acid is revealed by calculating R_f values. Each amino acid has a unique R_f value under particular conditions.

$$R_f = \frac{\text{distance moved by the amino acid}}{\text{distance moved by the solvent}}$$

Polypeptides

Amino acid monomers are joined together by **condensation reactions**. A **peptide bond** is formed between two amino acids, producing a **dipeptide**. A **polypeptide** is formed when many amino acids bond together. This occurs at ribosomes during **translation**. Peptide bonds are broken in hydrolysis reactions.

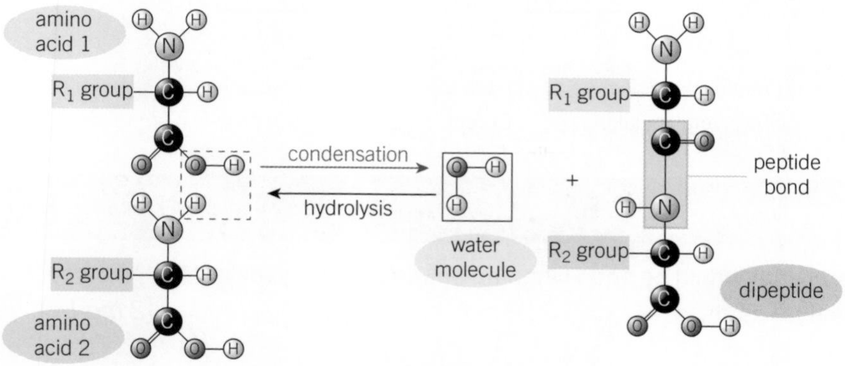

▲ **Figure 2** *Dipeptide synthesis and breakdown*

Levels of protein structure

A polypeptide produced during translation is called the **primary structure** of a protein. However, proteins are more complex than a sequence of amino acids in a chain. They fold and coil into shapes specific to their functions; these are called **secondary** and **tertiary structures**.

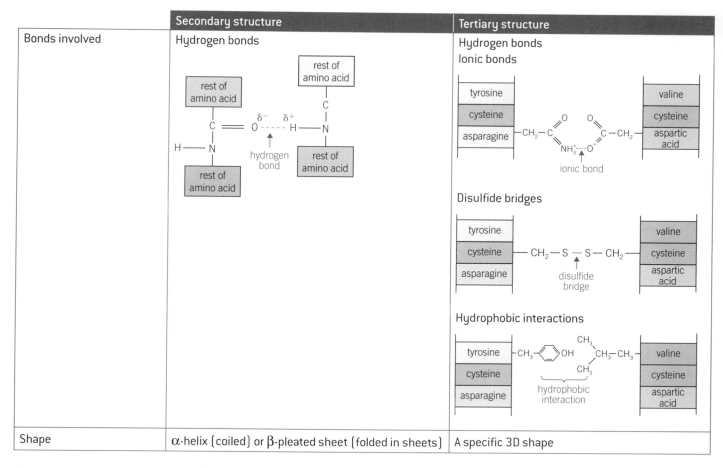

<table>
<tr><th></th><th>Secondary structure</th><th>Tertiary structure</th></tr>
<tr><td>Bonds involved</td><td>Hydrogen bonds
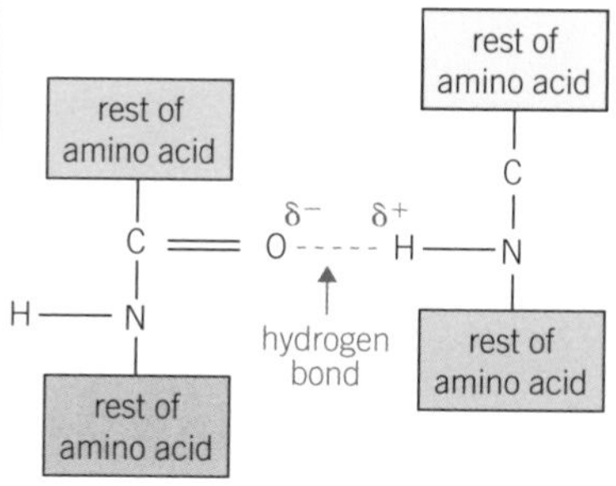</td><td>Hydrogen bonds
Ionic bonds
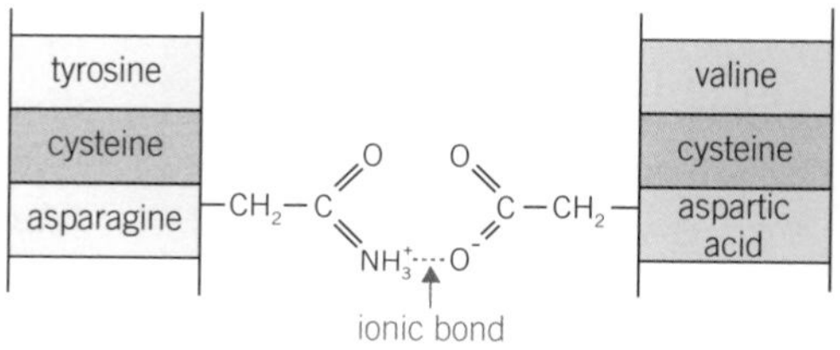
Disulfide bridges
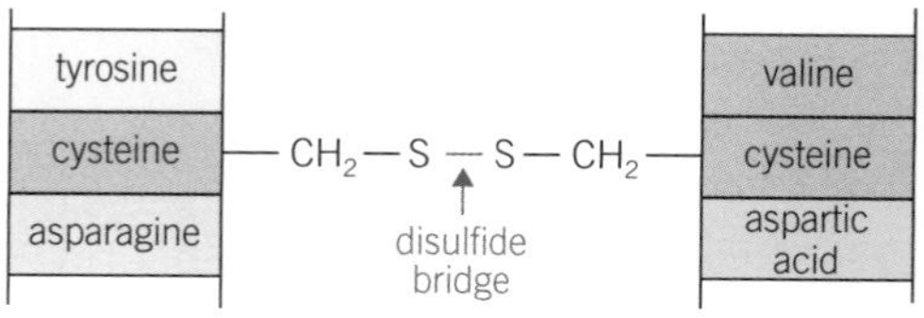
Hydrophobic interactions
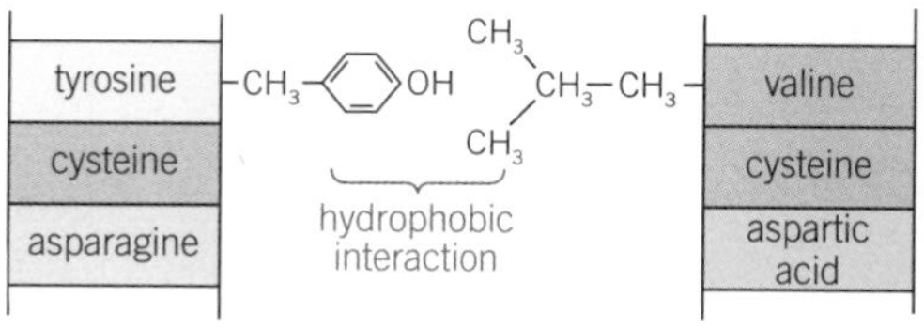</td></tr>
<tr><td>Shape</td><td>α-helix (coiled) or β-pleated sheet (folded in sheets)</td><td>A specific 3D shape</td></tr>
</table>

Quaternary structures are produced when two or more polypeptides associate. For example, haemoglobin consists of four polypeptide subunits and collagen comprises three polypeptides. Prosthetic groups (see Topic 4.4, Cofactors, coenzymes, and prosthetic groups) can be present in quaternary structures.

Common misconception: Polypeptide vs protein

A protein is more than a chain of amino acids. The polypeptide that is translated at ribosomes is a linear sequence of amino acids (the primary structure). Proteins, however, exhibit specific, complex shapes (due to bonds in their secondary and tertiary structures) and can contain several polypeptides.

Go further: Gluten

Fans of *The Great British Bake Off* hear gluten mentioned on a regular basis. Gluten is a protein that is pivotal to baking bread. Except … it's not just one protein. Gluten is a mixture of two types of protein: gliadins (which are smaller) and glutenins (which are larger).

Dough is kneaded to stretch gluten molecules. This allows more intermolecular bonds (e.g. hydrogen bonds) to form between glutenin and gliadin polypeptides, making the dough stronger and trapping CO_2 under gluten sheets.

1 Gluten proteins are insoluble in water. What does this suggest about the nature of the majority of their amino acids?

2 Explain why too much kneading of dough can create chewy bread.

3 Salt is added to dough to strengthen it by allowing gluten molecules to move closer to each other. Suggest how salt enables glutenins and gliadins to move closer together.

Summary questions

1 Draw alanine, which is an amino acid that has CH_3 as its R group. *(2 marks)*

2 Two amino acids were analysed using thin layer chromatography. The solvent was allowed to travel 5 cm along the gel. Calculate **a** the R_f value of amino acid X, which travelled 4.15 cm. **b** the distance travelled by amino acid Y, which has an R_f value of 0.71. *(2 marks)*

3 Suggest how the biuret test can be used to assess the concentration of proteins. *(4 marks)*

3.7 Types of proteins

Specification reference: 2.1.2(n) and (o)

In Topic 3.6, Structure of proteins, you learned how the tertiary and quaternary structures of proteins are produced. Proteins can be divided into two broad categories, based on the nature of these structures: **globular** and **fibrous** proteins. Here we compare the structures and properties of these two groups of proteins.

Synoptic link

We will explore the role of haemoglobin further in Topic 8.4, Transport of oxygen and carbon dioxide in the blood.

Synoptic link

You will learn more about prosthetic groups in Topic 4.4, Cofactors, coenzymes, and prosthetic groups.

Revision tip: Conjugated proteins

Some globular proteins are conjugated, which means they contain a non-protein component called a prosthetic group. Prosthetic groups can be metal ions, lipids, carbohydrates, or molecules derived from vitamins.

Globular and fibrous proteins

▼ **Table 1** *A comparison of globular and fibrous proteins*

	Globular	Fibrous
Shape	Compact and spherical	Long and linear
Bonding/structure	Hydrophilic R groups on the outside, and hydrophobic R groups on the inside	A limited range of amino acids, often with a repetitive sequence Organised and strong structures
Water solubility	Soluble	Insoluble
Conjugation (i.e. presence of prosthetic group)	Sometimes	No
Functions	Enzymes (e.g. catalase) Hormones (e.g. insulin) Membrane proteins Antibodies Transport proteins (e.g. haemoglobin, which has four polypeptides in its quaternary structure, each carrying a haem prosthetic group)	Structural roles For example: Keratin (in skin, nails, and hair) Collagen (in connective tissue in tendons, skin, and ligaments)

Summary questions

1 In which organelle are prosthetic groups added to polypeptides? *(1 mark)*

2 Explain, in terms of their functions, why it is important for globular proteins to be soluble and for fibrous proteins to be insoluble. *(2 marks)*

3 Keratin contains many disulfide bonds. Keratin in hair contains fewer disulfide bonds than keratin in nails. Suggest why. *(2 marks)*

3.8 Nucleic acids

Specification reference: 2.1.3(a), (b), and (d)(i), and (ii)

You have learned about two types of polymer (polysaccharides and polypeptides) in earlier topics. Here we introduce another class of polymer, nucleic acids, which are built from nucleotide monomers.

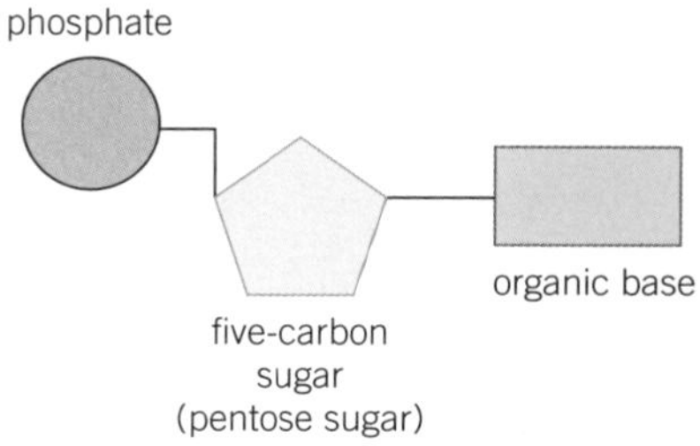

▲ **Figure 1** *The generalised structure of a nucleotide*

Nucleotide structure

A nucleotide has three components: a pentose (5-carbon) sugar, a phosphate group, and an organic nitrogenous base.

Nucleic acid polymers

Deoxyribonucleic acid (DNA) and **ribonucleic acid (RNA)** are both polynucleotides. You will discover more about their roles in the next couple of topics. Polynucleotides are formed by **condensation reactions** between nucleotides, which form **phosphodiester bonds**. As with other polymers you have encountered, polynucleotides are broken by **hydrolysis**.

Synoptic link

In Topic 3.11, ATP, we will discuss the importance of ATP as a phosphorylated nucleotide.

DNA

The sugar in DNA nucleotides is deoxyribose. Each DNA nucleotide has one of four bases: adenine (A), thymine (T), cytosine (C), or guanine (G).

Complementary base pairing

DNA exists as a double helix (i.e. two polynucleotides, running in opposite directions, bonded together). The two DNA polynucleotide strands are held together by hydrogen bonds between their bases. C on one strand bonds with G on the other strand; similarly, A always bonds with T.

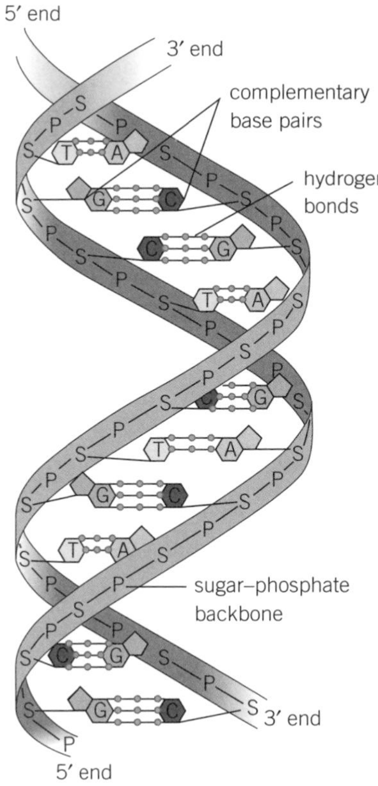

▲ **Figure 2** *DNA double helix – complementary base pairing between antiparallel strands*

▼ **Table 1** *Complementary base pairs in DNA*

Purine base (double-ring structure)	Number of hydrogen bonds between the bases	Pyrimidine base (single-ring structure)
 **Adenine**	Two hydrogen bonds between A and T	 **Thymine**
 **Guanine**	Three hydrogen bonds between G and C	 **Cytosine**

Revision tip: Basic memory tricks

A mnemonic is a memory tool. Mnemonics can sharpen memories by making them more visual. For example, *small pyramid CiTies; big, pure GArdens* may help you remember the relative size of DNA bases, and whether they are purines or pyrimidines.

Synoptic link

You learned about the structure of ribose in Topic 3.3, Carbohydrates.

RNA

RNA exists in three forms (messenger, transfer, and ribosomal RNA), all of which play roles in polypeptide synthesis. You will learn more about RNA in Topic 3.10, Protein synthesis. The structure of RNA is different to DNA in the following ways:

- The pentose sugar in RNA is **ribose** (rather than the deoxyribose in DNA).
- **Uracil** (U) is present instead of thymine; U bonds with A in RNA.
- RNA is **single-stranded** and does *not* form a double helix.

Practical skill: DNA purification

You can use the following steps to extract DNA from plant tissue:

1. **Grind** tissue (to break down cell walls)
2. Mix with **detergent** (to break down membranes)
3. Add **salt** (to break hydrogen bonds between DNA polynucleotides)
4. Add **protease** enzyme (to break down proteins surrounding DNA)
5. Add **alcohol** (to precipitate DNA out of solution)
6. Remove DNA (which can be seen as white strands below the alcohol layer).

Summary questions

1 Complete the table below, which features monomers and polymers you have encountered in this chapter. *(3 marks)*

Monomer	Polymer	Bond formed in condensation reaction
Monosaccharide		
	Polypeptide	
Nucleotide		

2 Explain why the DNA of a cell always contains equal amounts of adenine and thymine, and equal amounts of cytosine and guanine. *(2 marks)*

3 Suggest why C–A and T–G base pairings do not occur. *(2 marks)*

3.9 DNA replication and the genetic code

Specification reference: 2.1.3(e) and (f)

In Topic 3.8, Nucleic acids, you looked at the structure of DNA. Genetic material must be copied whenever a cell divides. In this topic we examine how DNA replication is achieved. You will also learn that the genetic code is degenerate and works through sequences of triplets.

DNA replication

The following stages occur in DNA replication:

1. Histone proteins are removed
2. The DNA double helix is **unwound** (by an enzyme called ***helicase***)
3. Hydrogen bonds between strands are broken (**unzipping**; catalysed by ***helicase***)
4. Both DNA strands act as **template strands**
5. Free (monomer) nucleotides are activated (phosphate groups are added to them)
6. Free nucleotides form hydrogen bonds with bases on the template strands
7. C bonds with G, A bonds with T (**complementary base pairing**)
8. Phosphodiester bonds join nucleotides in the new strands (catalysed by ***DNA polymerase***).

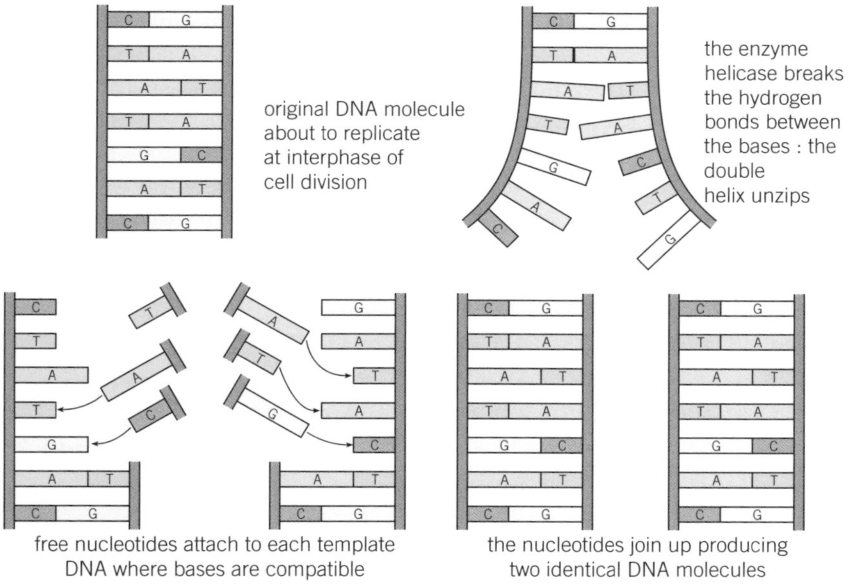

▲ **Figure 1** *Semi-conservative DNA replication*

The genetic code

The genetic code has the following properties:

- It uses a **triplet code** (i.e. a sequence of three nucleotides, known as a **codon**, codes for the production of one amino acid).
- The code is **degenerate** (i.e. in most cases, several codons code for the same amino acid). For example, TCT, TCC, TCA, TCG, AGT, and AGC all code for serine.

Revision tip: Why is DNA replication semi-conservative?

DNA replication results in two identical double helices (i.e. genetic information is conserved). In each helix, one strand remains from the original helix and one strand is new (hence *semi*-conservative).

Synoptic link

In Chapter 6, Cell division, you will look at the timing of DNA replication during the cell cycle and learn about cell division.

Revision tip: Degenerate maths

There are 64 (4^3) possible DNA codons. One of these is a start codon (which indicates the beginning of a gene sequence) and three are stop codons (which pinpoint the end of a gene). The remaining 60 codons are able to code for 20 different amino acids.

Summary questions

1. The base sequence GCCAAATCT appears in one strand of DNA. State the bases that would be present in the opposite strand. *(3 marks)*
2. Outline the importance of enzymes in DNA replication. *(4 marks)*
3. Explain why a genetic mutation (a change in the DNA sequence of a gene) does not always result in a different protein being produced. *(2 marks)*

3.10 Protein synthesis

Specification reference: 2.1.3(g)

Key terms

Transcription: The production of an mRNA molecule from a DNA base sequence (i.e. a gene).

Translation: The production of a polypeptide at ribosomes; the sequence of amino acids is determined by the codons in mRNA.

Synoptic link

Topic 2.4, Eukaryotic cell structure, provided an outline of the organelles involved in protein synthesis.

Revision tip: Replication vs transcription

The start of DNA replication and transcription are similar; DNA is unwound and hydrogen bonds between complementary bases are broken. However, the two processes show many differences. For example, during transcription only a small section of DNA is unzipped, only one strand acts as a template, RNA polymerase catalyses the reaction rather than DNA polymerase, and mRNA is produced rather than a new DNA strand.

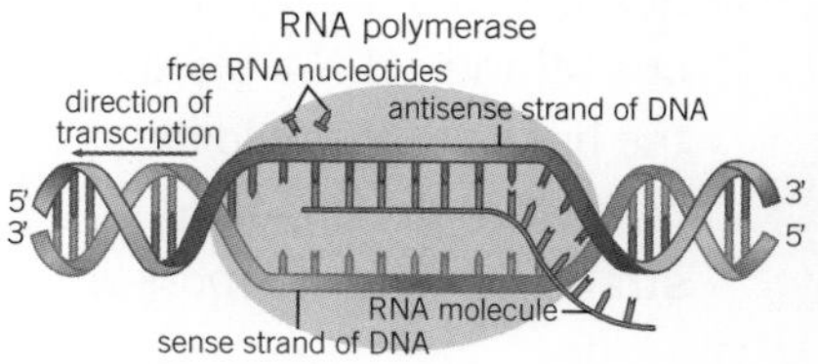

▲ **Figure 1** *Transcription: the role of RNA polymerase*

Two processes are needed to convert a genetic blueprint into a polypeptide: **transcription** and **translation**.

Transcription

Transcription produces a messenger RNA (mRNA) strand from a DNA sequence (a gene) using these steps:

1 ***Helicase*** **unwinds** and **unzips** DNA (but usually only along the base sequence of one gene).
2 **RNA nucleotides** bind to complementary DNA bases on the **template (antisense)** strand.
3 ***RNA polymerase*** joins RNA nucleotides together with phosphodiester bonds.
4 A **stop codon** causes RNA polymerase to detach.
5 The mRNA molecule detaches from the DNA and leaves the nucleus.

Translation

Translation occurs at ribosomes and involves the following stages:

1 **mRNA attaches** to a ribosome (in a groove between the two subunits).
2 The first mRNA **codon** binds to a **tRNA** molecule with a complementary **anticodon**.
3 The tRNA is bound to an amino acid specific to its anticodon.
4 A second tRNA (carrying another amino acid) binds to the adjacent mRNA codon.
5 A **peptide bond** forms between the two amino acids.
6 The ribosome continues along the mRNA molecule, from codon to codon.
7 Amino acids are bonded together in a polypeptide chain until a **stop codon** is reached.
8 The polypeptide is released.

Summary questions

1 Which DNA strand acts as the template strand during transcription? *(1 mark)*

2 Explain why DNA must be double-stranded despite only one strand representing the genetic code. *(2 marks)*

3 Complete Table 1, which compares DNA replication, transcription, and translation. *(7 marks)*

▼ **Table 1**	DNA replication	Transcription	Translation
Hydrogen bonds between complementary DNA base pairs are broken			
Free DNA nucleotides are activated			
To which base does adenine bond?			
Phosphodiester bonds are formed			
Peptide bonds are formed			
Location			
Product			

3.11 ATP

Specification reference: 2.1.2(b) and 2.1.3(c)

You were introduced to the structure of nucleotides in Topic 3.8, Nucleic acids. Adenosine triphosphate (ATP) is a phosphorylated nucleotide. It is the currency of chemical energy found in all organisms. In this topic you will learn about the structure of ATP and why it is suited to energy transfer.

ATP structure

ATP is similar in structure to an RNA nucleotide. It contains an adenine base and a ribose sugar. ATP, however, has three phosphate groups rather than the one found in an RNA nucleotide.

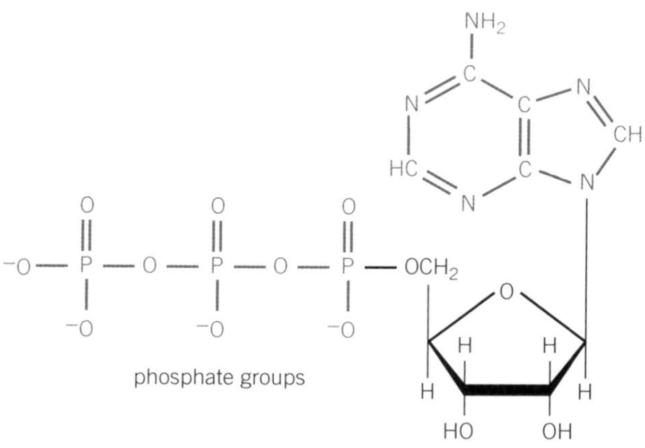

▲ **Figure 1** *ATP structure*

Revision tip: How does ATP deliver energy?

ATP releases energy to enable a vast range of reactions to occur. How is the energy transferred? A relatively small amount of energy is needed to remove a phosphate from ATP. Much more energy is released when the phosphate forms a new bond with another substance. Overall, energy is liberated and can be used in other metabolic processes.

Condensation and hydrolysis

ATP is formed in respiration through a condensation reaction. Hydrolysis of ATP produces ADP by removing one of the phosphates. This hydrolysis reaction releases energy that can be used in metabolic processes.

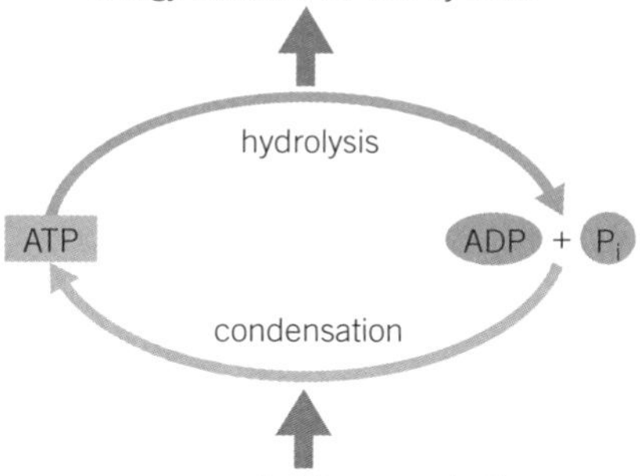

▲ **Figure 2** *The cycle of ATP production and use*

How is ATP suited to energy transfer?

The structure of ATP makes it an excellent energy carrier because each molecule:

- is soluble
- releases energy in small quantities (which prevents energy wastage)
- has an unstable phosphate bond (which is easily broken)

Summary questions

1. Outline the similarities and differences between RNA nucleotides and ATP. *(4 marks)*
2. Explain why ATP is not a suitable long-term energy store. *(2 marks)*
3. ATP is described as a universal energy carrier. Suggest the meaning of the term 'universal energy carrier'. *(2 marks)*

Chapter 3 Practice questions

1 Which of the following structural properties is possessed by glycogen?

A Contains only 1,4 glycosidic bonds

B Branched

C Formed from β-glucose

D Formed from monomers with the formula $C_5H_{10}O_5$ *(1 mark)*

2 Which of the following statements is a correct step in the emulsion test for lipids?

A Mix the sample with HCl

B Mix the sample with ethanoic acid

C Heat the mixture

D Shake the mixture *(1 mark)*

3 Which of the following bonds is the weakest of the bonds that hold the tertiary structure of proteins together?

A Hydrogen **C** Ionic

B Disulfide **D** Peptide *(1 mark)*

4 Which of the following statements is true of fibrous proteins?

A They can be conjugated

B Their amino acid sequences can be repetitive

C They are soluble

D Antibodies are examples of fibrous proteins *(1 mark)*

5 The following passage describes the construction of a calibration curve when using colorimetry to measure the concentration of fructose. Complete the passage by choosing the most appropriate word to place in each gap.

A .... filter is placed in the colorimeter, which is calibrated using a .... of distilled water. A calibration curve is obtained by performing the .... test on solutions with known fructose concentrations. The .... is filtered and colorimeter readings are obtained for each solution. *(4 marks)*

6 **a** Phosphorus is an element that is essential for plant growth. Describe how phosphorus is absorbed by plants. *(3 marks)*

b Outline the biological roles of phosphorus. *(6 marks)*

7 **a** Explain the role of water in temperature regulation in organisms.

(4 marks)

b Draw a diagram to show hydrogen bonding between a water molecule and an ethanol (CH_3CH_2OH) molecule. *(2 marks)*

8 **a** Explain the purpose of adding the following substances in the extraction of DNA from plant material. *(4 marks)*

- detergent • salt • protease • alcohol

b These questions concern the DNA base sequence ACGTTA.

i Describe the sequence in terms of purines and pyrimidines. *(2 marks)*

ii State the number of amino acids coded for by this sequence. *(1 mark)*

iii State the mRNA base sequence that would be transcribed by this DNA base sequence. *(2 marks)*

4.1 Enzyme action

Specification reference: 2.1.4(a), (b), and (c)

Many globular proteins are enzymes, which are biological catalysts.

The importance of enzymes

Metabolism is the term given to the sum of all the chemical reactions occurring in an organism. Enzymes control these metabolic reactions, which can either be **anabolic** (the formation of molecules from smaller units) or **catabolic** (the breaking down of molecules).

An example of an anabolic reaction is DNA replication, which is controlled by the enzyme DNA polymerase (see Topic 3.9, DNA replication and the genetic code). Digestion is a catabolic process; for example, the enzyme maltase breaks down maltose into two molecules of glucose.

Enzymes can be either intracellular (working inside cells) or extracellular (working outside cells). Most extracellular enzymes are catabolic. In multicellular organisms, digestive enzymes are secreted from cells into the digestive system. Unicellular organisms, such as yeast and bacteria, secrete digestive enzymes into their immediate environment.

How do enzymes work?

All enzymes operate through the same principles:

1 **Substrates** collide with the **active site** of the enzyme.
2 The shape of the active site is **complementary** to the substrate.
3 The substrate binds to the active site to form an **enzyme–substrate complex** (ESC).
4 Bonds in the substrates are placed under strain and break; the enzyme provides an alternative reaction pathway that reduces the **activation energy** required for the reaction.
5 An **enzyme–product complex** is formed and the product(s) is/are released.

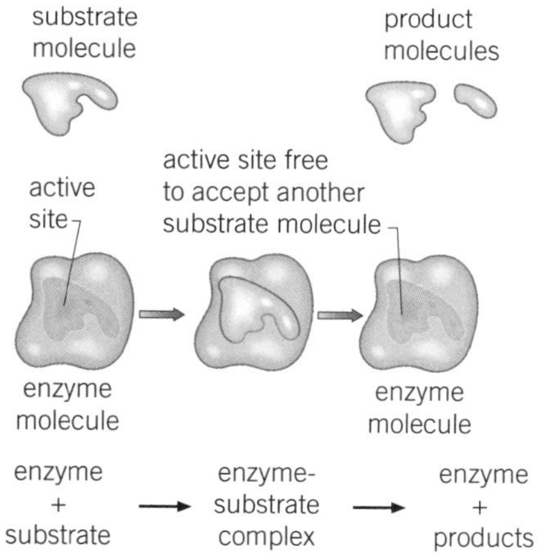

▲ **Figure 1** *The mechanism of enzyme action (illustrated here by a catabolic reaction)*

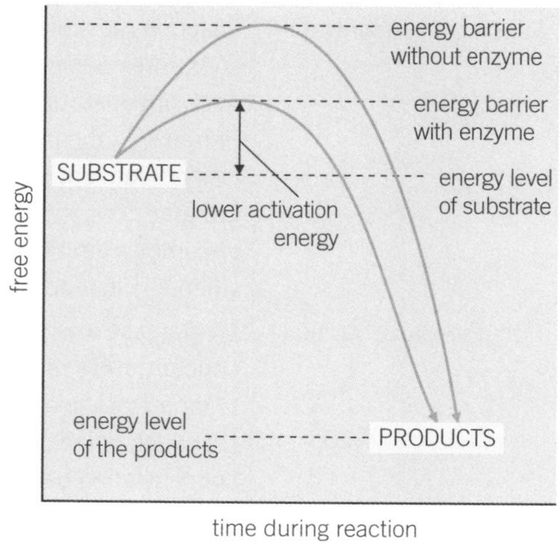

▲ **Figure 2** *Enzymes lower the activation energy of a reaction (i.e. less energy is required to initiate the reaction)*

Two competing theories exist to explain the enzyme–substrate binding mechanism. The **lock and key** hypothesis suggests that the shape of the active site is an ideal fit for the substrate molecule and is therefore **specific** to one substrate. The **induced-fit** hypothesis, however, suggests that initially weak binding by the substrate will alter the enzyme's tertiary structure. This strengthens the temporary bonds between the substrate and the enzyme, and weakens bonds within the substrate.

Key term

Enzymes: Biological catalysts that facilitate chemical reactions.

Synoptic link

You learned about globular proteins in Topic 3.7, Types of proteins.

Key terms

Substrate: A reactant in an enzyme-catalysed reaction.

Active site: The region of an enzyme to which substrates bind.

Revision tip: Enzyme names

Enzymes often end with the suffix '-ase' and are named after the substrate on which they act. Lactase, for example, breaks down lactose, and glycogen synthase catalyses the synthesis of glycogen. Some enzymes, however, do not conform to this naming system (e.g. trypsin breaks down polypeptides into smaller peptides).

Summary questions

1 State three structural features that are common to all enzymes. *(3 marks)*

2 State whether the following reactions are anabolic or catabolic. Explain your answers.
a The formation of glycogen **b** The formation of maltose from amylose **c** Glycerol formation from a triglyceride *(3 marks)*

3 Explain how enzymes are able to lower the activation energy of reactions. *(4 marks)*

4.2 Factors affecting enzyme activity

Specification reference: 2.1.4(d)

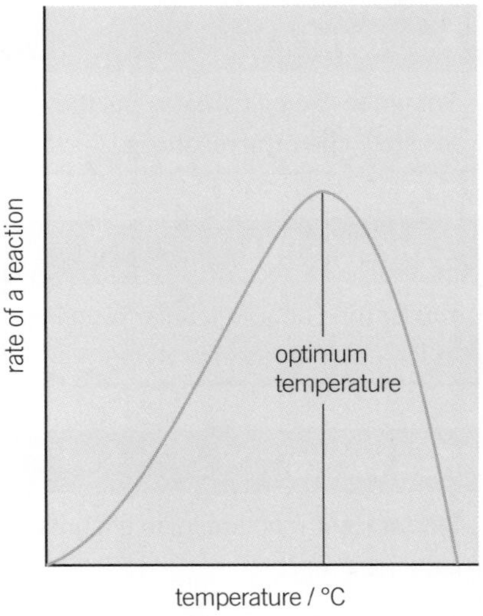

▲ **Figure 1** *All enzymes have an optimum temperature at which they work*

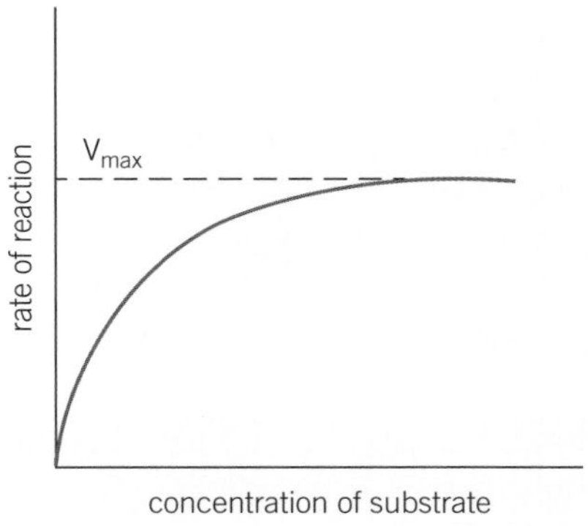

▲ **Figure 2** *Reaction rates increase as substrate concentration increases until V_{max} is reached*

You learned about how enzymes facilitate biological reactions in Topic 4.1, Enzyme action. An enzyme's performance, however, is dependent on several factors, which we discuss here.

Which factors affect enzyme activity?

The following table shows the principal factors affecting the performance of enzymes.

Factor	Effect	Explanation
Temperature	A rise in temperature will increase enzyme activity up to an optimum temperature	Enzyme and substrate molecules gain kinetic energy and move faster, thereby increasing the chance of successful collisions
	Above the optimum temperature, further increases in temperature reduce and eventually stop enzyme activity	Weak bonds (e.g. hydrogen bonds) in the active site vibrate more, strain, and break (i.e. the enzyme **denatures**)
pH	Enzyme activity decreases when pH moves away from the optimum (either decreasing or increasing)	Changes to H^+ ion concentrations alter the charges on amino acids in the active site, which can prevent substrate molecules from binding
	A further change in pH can stop all enzyme activity	Hydrogen and ionic bonds in the active site are broken, which causes a permanent change in the enzyme's tertiary structure (i.e. the enzyme is denatured)
Substrate concentration	Reaction rate increases as substrate concentration rises but eventually plateaus (when V_{max}, the maximum rate of reaction, is reached). Enzyme concentration becomes a limiting factor when V_{max} is reached.	A higher collision rate between substrates and active sites results in enzyme–substrate complexes (ESCs) forming at a greater rate. Reaction rate plateaus when one of the factors becomes limiting.
Enzyme concentration	An increase in enzyme concentration will raise the reaction rate to a higher V_{max} (at which point substrate concentration becomes the limiting factor)	

Revision tip: One enzyme's optimum is another enzyme's denaturation

Optimum temperatures and pH vary between enzymes, depending on where they operate. For example, enzymes in the human body tend to function best at approximately 37.5 °C, whereas thermophilic bacteria possess enzymes with much higher optimum temperatures. Digestive enzymes operating in the acidic environment of the stomach have a low optimum pH (e.g. pH 2), whereas the higher pH of the small intestine requires a higher optimum pH (e.g. pH 7).

Common misconception: What counts as denaturation?

All proteins, not only enzymes, can be denatured. A protein is denatured when bonds that hold its secondary or tertiary structure together are broken. These bonds might include hydrogen bonds, ionic bonds, and disulfide bridges. Bonds can be broken by temperature increases (above an optimum) and large shifts in pH. An alteration of secondary or tertiary structure usually results in a loss of function.

The following are not examples of denaturation:

- a small decrease in temperature (which slows enzyme activity due to a reduction of kinetic energy); exposure to very cold temperatures, however, can result in cold denaturation of enzymes, which you are not required to understand.
- enzyme inhibition (see Topic 4.3, Enzyme inhibitors)
- the breaking of peptide bonds in a protein's primary structure.

Practical skills: How to investigate factors that affect enzyme activity

When investigating enzyme activity, several factors can act as independent variables (e.g. temperature, pH, substrate concentration, enzyme concentration, and cofactor concentration – see Topic 4.4, Cofactors, coenzymes, and prosthetic groups). Remember, in a well-designed experiment only the independent variable will change. All other factors must be controlled. You will therefore be testing the effect of a single factor (the **independent variable**) on the rate of an enzyme-catalysed reaction (the **dependent variable**).

The values used in the investigation should be appropriate (i.e. within realistic ranges) if the results are to be **valid**.

Repetitions of the experiment should be conducted to improve the **reproducibility** of the results.

Example: investigating the effect of temperature on the activity of lipase

Independent variable = temperature (a water bath can be set to a range of temperatures, e.g. 0–60 °C)

Dependent variable = the rate at which lipids are broken down (for example, the conversion of triglycerides in milk to fatty acids can be monitored using a pH indicator)

Factors to control (which should be the same for each temperature tested) = the initial pH, lipase concentration, the volume of milk, the overall volume of solution.

Summary questions

1 Explain how high temperatures and changes in pH can prevent enzymes from functioning. *(5 marks)*

2 Suggest why microorganisms that inhabit alkaline lakes are unlikely to cause infections in humans. *(2 marks)*

3 Psychrophiles are organisms that inhabit cold environments where temperatures can drop below 0 °C. Suggest how the structure of psychrophiles' enzymes might differ from those found in humans. *(2 marks)*

Key term

Denaturation: The (usually permanent) change in the tertiary structure of a protein.

Revision tip: The temperature coefficient

When assessing the effect of temperature on the rate of enzyme-controlled reactions, a good rule of thumb is that a 10 °C rise in temperature will double the reaction rate. In other words, Q_{10} (the temperature coefficient) is 2.

4.3 Enzyme inhibitors

Specification reference: 2.1.4(f)

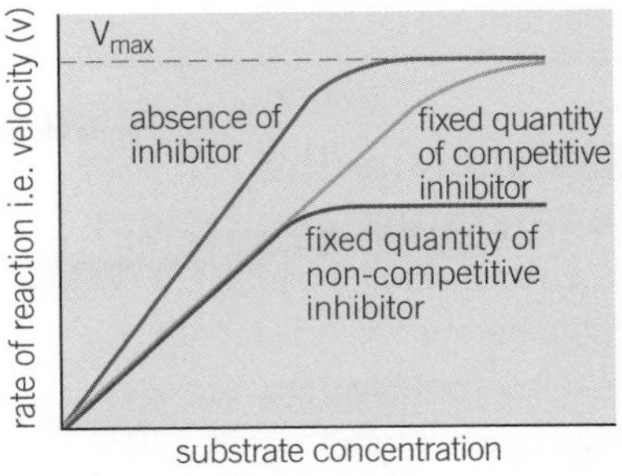

▲ **Figure 1** *The effect of substrate concentration on reaction rate in the presence and absence of inhibitors*

Common misconception: Competitive inhibitors – similar but not the same

Take care when answering questions about competitive inhibitors – the wording is important. A competitive inhibitor has a *complementary* shape to the active site of the enzyme (not the *same* shape). We can therefore conclude that the inhibitor has a *similar* shape to the substrate (but not the *same* shape, overall). The inhibitor and the substrate might, however, share some of the same groups.

Summary questions

1. State two differences in the mechanisms exhibited by competitive and non-competitive inhibitors. *(2 marks)*
2. Sulfanilamide is an antibiotic. It inhibits a bacterial enzyme that helps to convert a molecule called PABA into folic acid. PABA and sulfanilamide have different structures, but both contain a benzene ring and an amine group. Suggest how sulfanilamide inhibits bacterial enzymes. *(4 marks)*
3. Explain why end-product inhibition is an example of negative feedback. *(2 marks)*

In Topic 4.2, Factors affecting enzyme activity, you examined how enzyme activity can be limited by factors such as temperature, pH, and the concentrations of enzyme and substrate molecules. Here we consider how certain chemicals, known as inhibitors, can bind to enzymes, thereby slowing the rate of the reactions they catalyse.

Competitive and non-competitive inhibition

Enzyme inhibitors can be either competitive or non-competitive. The following table provides a summary of the features of these two forms of inhibition.

	Competitive	Non-competitive
Binding site	Active site	Allosteric site (i.e. a region of the enzyme that is not the active site)
Mechanism	Competes with the substrate for the enzyme's active site and blocks the substrate from entering	Tertiary structure of the enzyme (and therefore active site) changes shape, meaning the active site is no longer a complementary shape to the substrate
Reversible or irreversible	Reversible (usually)	Sometimes reversible, sometimes not
Examples	*Penicillin*: an antibiotic that inhibits the active site of the bacterial enzyme transpeptidase, thereby preventing cell wall formation *Statins*: inhibit HMG-CoA reductase (an enzyme involved in cholesterol production)	*Cyanide ions (CN^-)*: a metabolic poison that inhibits cytochrome c oxidase (a respiratory enzyme) *Organophosphates*: they are used as insecticides; these chemicals inhibit acetylcholinesterase, which is an enzyme necessary for the correct functioning of insect (and mammalian) nervous systems
Effect on reaction rate	Slows rate, but V_{max} (maximum rate) can still be reached if substrate concentration is increased	Slows rate and lowers V_{max}
What does it look like?	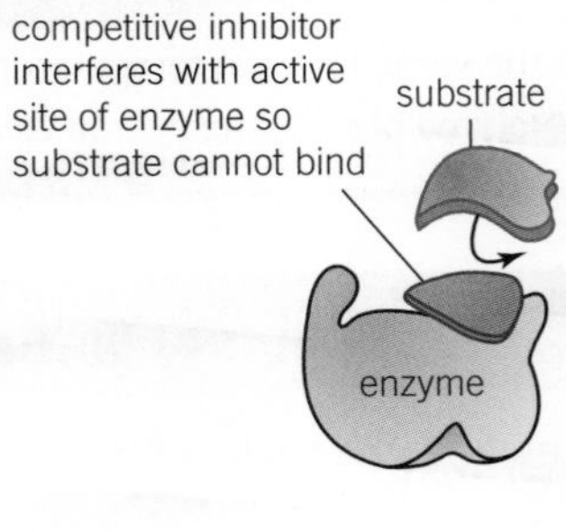	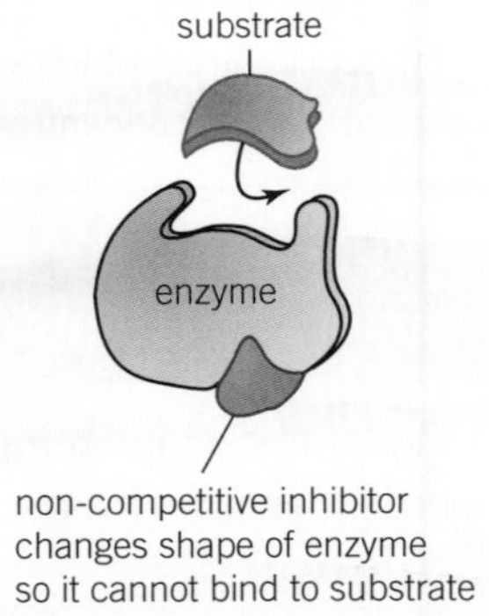

End-product inhibition

End-product inhibition occurs at the conclusion of metabolic reaction pathways. The final product in a series of reactions will often inhibit one of the enzymes in the reaction pathway from which it has been produced. This regulates the rate at which the product is made (i.e. it acts as a negative feedback mechanism). End-product inhibition can be competitive or non-competitive.

4.4 Cofactors, coenzymes, and prosthetic groups

Specification reference: 2.1.4 (e) and (f)

In Topic 4.3, Enzyme inhibitors, you learned about substances that inhibit enzyme activity. Now you will examine substances that have the opposite effect and are required for enzymes to work. These substances are known as cofactors.

Categories of cofactors

Cofactors are non-protein substances that enable proteins, including enzymes, to function. They can be either organic or inorganic. Prosthetic groups comprise a subset of cofactors that are permanently attached to the enzymes they assist. Other cofactors form temporary bonds with their enzymes–if organic, they are known as coenzymes. Most coenzymes are derived from vitamins. The following table summarises the characteristics of the different types of cofactor.

	Coenzyme	Inorganic cofactor	Prosthetic group
Are they organic (i.e. containing carbon) or inorganic?	Organic	Inorganic	Can be organic or inorganic
How do they interact with enzymes?	They form temporary bonds with the enzyme but leave following the reaction		Permanently bound to the enzyme
Examples	*Coenzyme A* (enzyme = acetyl CoA carboxylase) See 5.2.2(f) *NAD*$^+$ (enzyme = lactate dehydrogenase) See 5.2.2(f)	*Cl*$^-$ ions (enzyme = amylase) See 2.1.2(f) for the structure of amylose *Mg*$^{2+}$ (enzyme = DNA polymerase) See 2.1.3(e)	*Zn*$^{2+}$ (enzyme = carbonic anhydrase) See 3.1.2(i) *FAD* (enzyme = succinate dehydrogenase, an enzyme used in respiration) See 5.2.2

Precursor activation

Often enzymes are produced as inactive precursor proteins. The addition of a cofactor can activate an enzyme. The activation is usually achieved by altering the shape of the enzyme's active site.

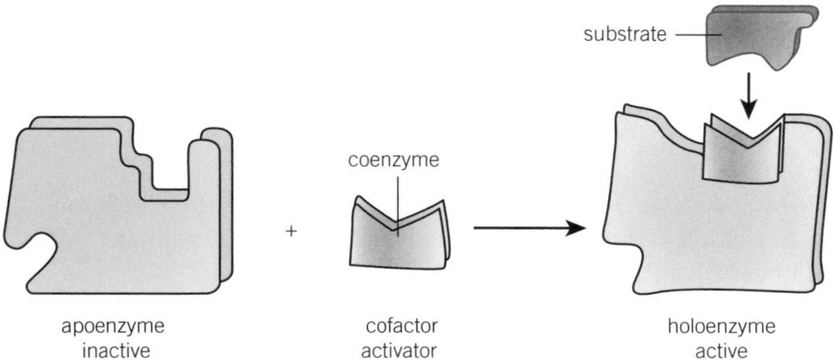

▲ **Figure 1** *A cofactor can modify the shape of an active site*

Synoptic link

You learned about prosthetic groups in Topic 3.6, Structure of proteins.

Key term

Cofactor: A non-protein substance required for enzymes to function.

Summary questions

1 Outline the differences between coenzymes and prosthetic groups. *(2 marks)*

2 Suggest why some digestive enzymes are produced as inactive proteins. *(2 marks)*

3 Using your knowledge from earlier chapters, suggest how polypeptides are modified to produce enzymes with prosthetic groups. *(4 marks)*

Chapter 4 Practice questions

1 Which of the following is most likely to result in the denaturation of a protein?

A The breaking of peptide bonds
B Competitive inhibition
C A small decrease in temperature
D A small increase in pH *(1 mark)*

2 A solution of amylase was diluted using the following method:

1 ml of the stock solution was added to 9 ml of distilled water (forming a second solution). 2 ml of the second solution was then transferred to 8 ml of distilled water (forming a third solution). For the final dilution, 1 ml of the third solution was added to 19 ml of distilled water (forming a fourth solution).

Which of the following represents the overall dilution from the stock solution of amylase to the fourth solution?

A 3-fold dilution
B 100-fold dilution
C 1000-fold dilution
D 5000-fold dilution *(1 mark)*

3 Which of the following statements is true of non-competitive inhibitors?

A They lower V_{max}
B Their effects are always reversible
C They bind to active sites
D They alter the primary structure of an enzyme *(1 mark)*

4 Which of the following statements is true of apoenzymes?

A They can be activated by the addition of a prosthetic group
B They are active precursors
C The shapes of their active sites can be altered by coenzymes
D They are formed from holoenzymes *(1 mark)*

5 A group of students investigated the effect of pH on the activity of amylase. Their results are shown in the following table.

pH	Relative activity
4	30%
4.5	45%
6	85%
7	100%
9	65%

a Based on the students' results, what is the optimum pH of amylase? *(1 mark)*

b State two improvements that the students could make to the presentation of their results. *(2 marks)*

c Suggest how the students could improve the validity of their results. *(1 mark)*

6 The following passage describes the mechanism of competitive enzyme inhibition. Complete the passage by choosing the most appropriate word to place in each gap.

A competitive inhibitor has a .... shape to the .... of an enzyme and a .... shape to the enzyme's substrate. *(3 marks)*

7 Suggest three factors that should be controlled when investigating the effect of substrate concentration on enzyme activity. *(3 marks)*

5.1 The structure and function of membranes

Specification reference: 2.1.5(a) and (b)

The organelles and molecules discussed in previous chapters need to be organised and partitioned within cells. This organisation is the role of membranes, which we explore in this topic.

Roles of membranes

Membranes are located at cell surfaces (i.e. plasma membranes) and surrounding organelles (in eukaryotic cells). They have the following functions:

- **Physical barriers** (they separate intracellular environments from extracellular environments)
- They regulate the **exchange of substances** in and out of a cell; membranes are *partially permeable* (i.e. they allow only certain particles to pass through)
- **Compartmentalisation** (i.e. membranes enclose and isolate organelles, enabling them to maintain specific environments for chemical reactions)
- Support for the **cytoskeleton**
- Sites of **chemical reactions**
- Sites of **cell communication** (e.g. cell signalling)
- Formation of spheres called **vesicles**, which are used in bulk transport (see Topic 5.4, Active transport).

Membrane structure

Membranes consist primarily of a phospholipid bilayer, which has a hydrophobic core and hydrophilic phosphate heads. Various proteins are distributed within the bilayer. Membrane structure can be visualised as a fluid mosaic model ('fluid' because phospholipids are able to move within the bilayer, and 'mosaic' because proteins are scattered throughout the bilayer, like the tiles in a mosaic).

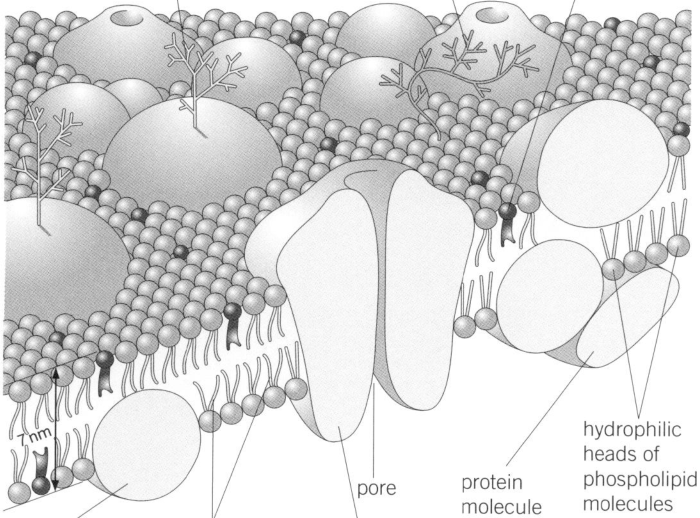

▲ **Figure 1** *The fluid mosaic model of membrane structure*

Synoptic link

You were introduced to phospholipids in Topic 3.5, Lipids.

▼ **Table 1** *Some of the components of membranes, in addition to phospholipids and cholesterol*

Component	Roles
Channel proteins	Facilitated diffusion (see Topic 5.3, Diffusion)
Carrier proteins	Facilitated diffusion and active transport (see Topics 5.3, Diffusion, and 5.4, Active transport)
Glycoproteins	Receptors (e.g. for neurotransmitters, peptide hormones, and drugs)
Glycolipids	Cell recognition (i.e. they act as antigens – see Chapter 12, Communicable diseases)

Go further: Cholesterol, a membrane buffer

Cholesterol is a lipid molecule that regulates the fluidity of membranes by inserting itself between phospholipids. The effect of the molecule, however, is temperature-dependent. At low temperatures cholesterol maintains fluidity by widening the gaps between phospholipids. Cholesterol prevents membranes becoming too fluid at high temperatures by attracting phospholipids and limiting their movement. Cholesterol therefore improves membrane stability.

1 Explain why cholesterol is sometimes referred to as a membrane 'buffer'.

2 Psychrophilic bacteria are adapted to cold habitats. Suggest the principal benefit of cholesterol in the plasma membranes of psychrophilic species.

Summary questions

1 Outline three roles of membranes *within* cells. (*3 marks*)

2 State and explain which type of protein within membranes would be used in the following cases: **a** The uptake of glucose into a cell **b** The binding of adrenaline to a liver cell. (*4 marks*)

3 Some cells, such as cells in the proximal convoluted tubules of the kidneys, require different proteins to be present on either side of the cell. Suggest why an increase in membrane fluidity would be a problem for these cells. (*2 marks*)

5.2 Factors affecting membrane structure

Specification reference: 2.1.5(c)

Membranes dictate the shape and content of cells and organelles. A loss of membrane structure would therefore damage the cell or organelle the membrane surrounds. Factors such as temperature and the presence of solvents determine the integrity of membrane structure.

Factors affecting membrane structure

Apart from shortages of membrane components, such as phospholipids and cholesterol, two factors can have a negative impact on membrane structure: temperature changes and the presence of solvents.

▼ **Table 1** *The effects of temperature and solvents on membrane structure*

Factor	Effect on membrane
Temperature	Decreased temperature reduces fluidity (phospholipids move less due to lower kinetic energy) Increased temperature causes greater fluidity and therefore increases permeability, but a membrane will lose its structure and break apart if temperature continues to rise
Solvents	Organic, weakly polar (e.g. ethanol) or non-polar (e.g. benzene) solvents disrupt or dissolve membranes, making them more permeable

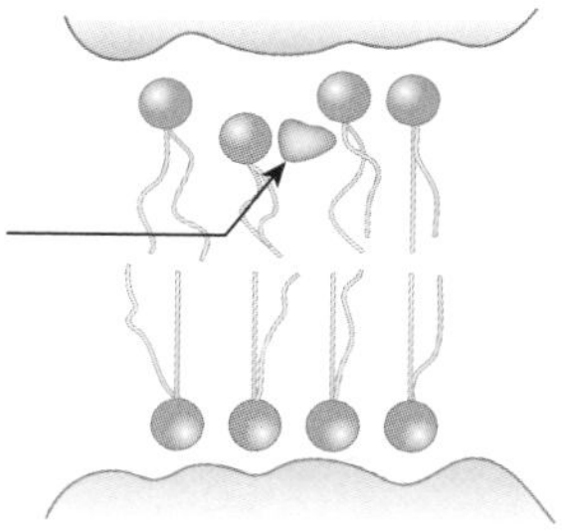

▲ **Figure 1** *Alcohol molecules (e.g. ethanol) disrupt membrane structure, thereby increasing permeability*

Revision tip: Membranes do not denature

Proteins within membranes can be denatured (see Topic 4.2, Factors affecting enzyme activity). Membranes as a whole (i.e. the phospholipid and cholesterol components) are not denatured by high temperatures. You can describe a membrane as being 'disrupted' or 'destroyed' by increases in temperature.

Practical skills: Investigating membrane permeability

Question and model answer

The effect of solvents, temperature, pH, and other factors on membrane permeability can be investigated experimentally.

The effect of alcohol, for example, can be studied using the following procedure:

Five test tubes of alcohol solution (0%, which is a control, 10%, 20%, 30%, and 40%) are set up. Beetroot cylinders are added to each tube. After 10 minutes, the colour of each alcohol solution is measured. A red pigment called betacyanin leaks from beet cells into the surrounding solution when membranes are damaged. The intensity of the colour released is proportional to the level of cellular damage.

1 State three factors the investigator should control in order to increase the validity of the experiment.
Answers include: temperature, volume of alcohol solution, surface area: volume ratio of beet cylinders, pH.

2 The procedure is repeated three times. Explain how this improves the experiment.
Answer: reproducibility is increased, and the spread of results can be assessed (using statistical tests).

3 Suggest how the colour of the alcohol solutions can be measured.
Answer: using a colorimeter.

Summary questions

1 Describe how temperature affects the permeability of membranes. *(3 marks)*

2 Other than the effect on phospholipids, why else might membrane function be damaged at high temperatures? *(2 marks)*

3 Explain how excessive ethanol consumption might disrupt membrane function in cells. *(5 marks)*

5.3 Diffusion

Specification reference: 2.1.5(d)

One of the functions of membranes, as you learned in Topic 5.1, The structure and function of membranes, is to determine which substances move in and out of cells and organelles. The next three topics are concerned with the methods by which such transmembrane movement is achieved. You will begin by learning about passive transport.

Diffusion across membranes

Diffusion is the net (overall) movement of particles from a region of higher concentration to a region of lower concentration. It is a passive process (i.e. not requiring energy from ATP).

Diffusion through the bilayer

Some particles can diffuse through the bilayer, between phospholipid molecules. Large lipid-soluble molecules (e.g. steroid hormones), non-polar molecules (e.g. oxygen), and very small polar molecules (e.g. water) are able to pass directly through the bilayer.

Facilitated diffusion

Ions and large polar molecules (e.g. glucose and amino acids) pass through proteins rather than between phospholipids. The process of particles passing through transmembrane proteins is called facilitated diffusion. Two types of proteins are used:

Channel proteins: pores, which can be gated (i.e. opened and closed), allowing the diffusion of ions.

Carrier proteins: they have shapes that allow only the passage of specific molecules or ions.

Factors affecting diffusion rates

Higher temperatures increase diffusion rate by providing particles with more kinetic energy. In addition, the rate of diffusion increases with the following membrane characteristics:

- Steeper **concentration gradient** – the greater the difference in concentration either side of a membrane, the greater the diffusion rate
- Shorter **diffusion pathway** – thin membranes reduce the distance particles have to move
- Greater **surface area**
- A higher **concentration of carrier proteins** increases the rate of facilitated diffusion.

Practical skill: Investigating factors that affect diffusion rates

Dialysis (Visking) tubing is a partially permeable material that can be used to simulate a cell membrane. This tubing is often used in experiments to test the effect of concentration gradients or temperature on diffusion rate.

Revision tip: Down, not across

Particles move *down* concentration gradients (from a high to low concentration) rather than *along* or *across* gradients.

Synoptic link

You will learn more about surface area to volume ratio and its influence on diffusion rate in Topic 7.1, Specialised exchange surfaces.

Summary questions

1 Glucose molecules are reabsorbed from proximal convoluted tubules in the kidney. Explain why the cell membranes of the proximal convoluted tubule: **a** are folded into microvilli; and **b** contain a high concentration of carrier proteins. (*4 marks*)

2 State how the following particles are likely to diffuse through a membrane. Explain your answers. **a** carbon dioxide **b** potassium ions. (*4 marks*)

3 Fick's law expresses diffusion rate in relation to surface area, membrane thickness, and concentration gradient. Complete the following equation, which represents a simplified version of Fick's law. Diffusion rate is proportional to __________ x __________ / __________. (*3 marks*)

5.4 Active transport

Specification reference: 2.1.5(d) (i)–(ii)

The movement of particles down a concentration gradient does not require energy. You learned about this type of passive movement in Topic 5.3, Diffusion. Here we discuss movement against a concentration gradient, which requires energy. These particle movements across membranes are referred to as active transport.

Active transport

Energy is required for active transport because particles are being moved against a concentration gradient (i.e. from a region of lower to a region of higher concentration). The energy is in the form of ATP. Active transport can use carrier proteins as pumps or take the form of bulk transport.

Carrier proteins (pumps)

Carrier proteins change shape, thereby allowing particles to pass through them, due to ATP hydrolysis (which produces phosphate ions).

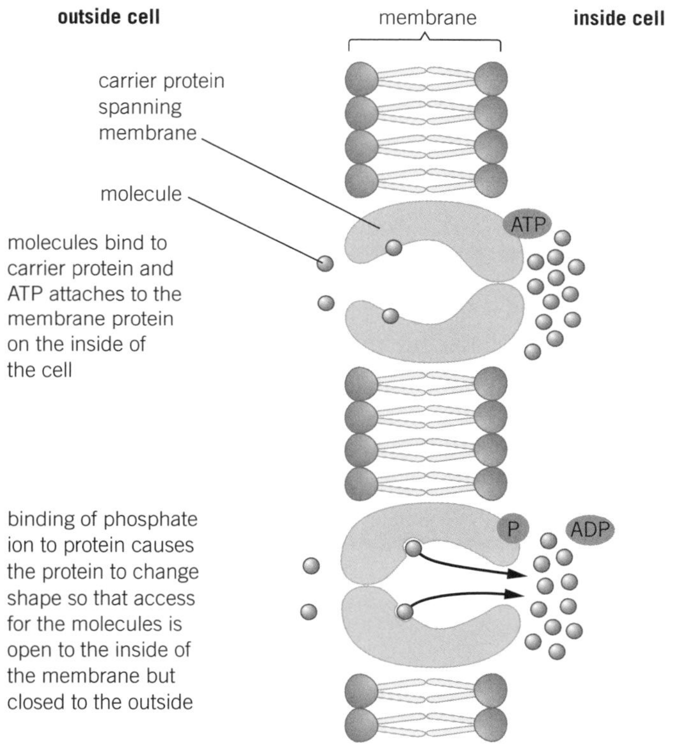

▲ **Figure 1** *The mechanism by which carrier proteins function*

Bulk transport

The movement of large molecules (e.g. enzymes and hormones) relies on bulk transport, which is the movement, in and out of cells, of particles within vesicles.

Endocytosis

Endocytosis is bulk transport *into* cells. Vesicles are formed by the plasma membrane being pinched off. Two forms exist:

Pinocytosis: a cell engulfs liquid and small dissolved particles

Phagocytosis: a cell engulfs large solid material (e.g. a white blood cell engulfing a bacterium).

Synoptic link

We discussed ATP as a molecule that transfers chemical energy in Topic 3.11, ATP.

Common misconception: Facilitated diffusion vs active transport

Carrier proteins are used for facilitated diffusion and active transport. In both processes, carrier proteins are selective (i.e. they change shape only when a specific particle binds to the carrier protein). Carrier proteins used for active transport, however, require ATP hydrolysis in addition to the binding of the particle. Facilitated diffusion does not require ATP.

Revision tip: Phagocytosis is not just for phagocytes

Phagocytosis is usually discussed in the context of phagocytes (cells of the immune system that take in and digest pathogens), but phagocytosis can occur in other cells.

Synoptic link

You will learn about the importance of phagocytosis as part of the immune system in Topic 12.5, Non-specific animal defences against pathogens.

Exocytosis

Exocytosis is bulk transport *out* of cells. In many cases, vesicles are formed by the Golgi apparatus and fuse with the cell surface membrane, releasing their contents (e.g. hormones).

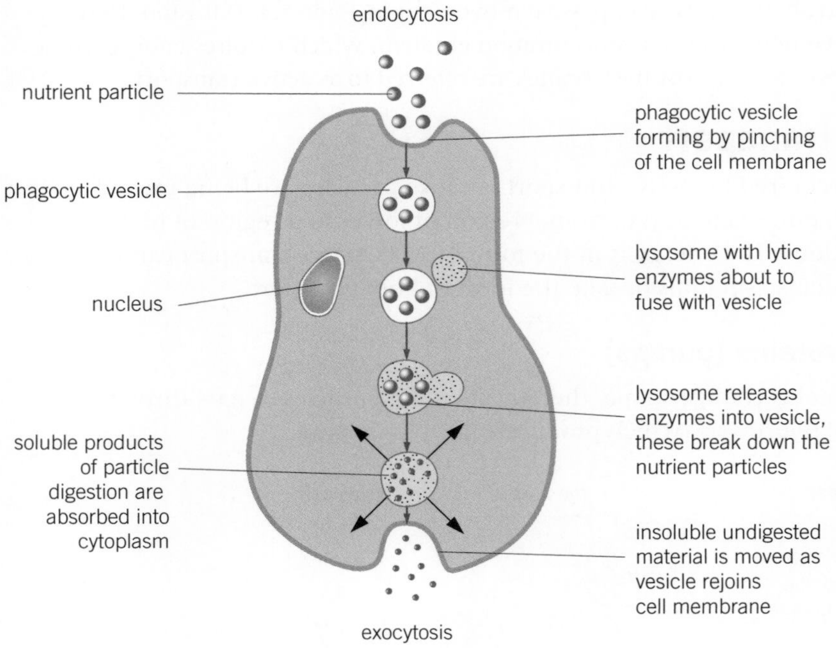

▲ **Figure 2** *An example of bulk transport*

Summary questions

1 Outline the characteristics of particles that are transported using exocytosis. *(4 marks)*

2 Complete the following table, which compares facilitated diffusion and active transport. *(5 marks)*

	Active transport	Facilitated diffusion
Uses carrier proteins		
Particles move down concentration gradients		
Particles move against concentration gradients		
Requires ATP		
At least two binding sites must be present on carrier proteins		

3 Suggest how the vesicles formed during phagocytosis would differ from those formed during pinocytosis. Explain your answer. *(3 marks)*

5.5 Osmosis

Specification reference: 2.1.5(e) (i)–(ii)

Osmosis and water potential

The direction in which water diffuses across a membrane is determined by water potential (ψ), which is measured in kilopascals (kPa). Water potential decreases as more solute particles dissolve. Osmosis always operates from a region of higher water potential to a region of lower water potential; in other words, water moves to the side of the membrane that has more dissolved solute.

▼ **Table 1** *Comparison of the water potentials of pure water and glucose solutions*

	Water potential	Solute concentration
Pure water	Highest possible water potential (0 kPa)	No solute dissolved
Dilute glucose solution	High water potential (−20 kPa)	Low glucose concentration
Concentrated glucose solution	Low water potential (−400 kPa)	High glucose concentration

Effects of osmosis on plant and animal cells

Osmosis produces different effects in plant and animal cells. These differences result from the cellulose walls surrounding plant cells.

▼ **Table 1** *The effects of osmosis on animal and plant cells*

Water potential of the solution surrounding the cell	Net movement of water	Effect on animal cell	Effect on plant cell
Higher (less solute in the surrounding solution)	Enters cell	Swells and bursts (i.e. the cell undergoes **lysis**)	Swells and becomes **turgid** (i.e. the cell is full of water and the membrane is pushed against the cell wall)
Equal (the same solute concentrations in the cell and surrounding solution)	Water leaves and enters the cell, but at equal rates	No change	No change
Lower (more solute in the surrounding solution)	Leaves cell	Shrinks (i.e. the cell undergoes **crenation**)	**Plasmolysis** (membrane pulls away from the cell wall)

Summary questions

1 State whether the following sentences are true or false. **a** Diluting a solution will lower water potential. **b** No solutions can have a water potential of 0. **c** A liver cell is in danger of plasmolysing if placed in a solution with a high water potential. *(3 marks)*

2 An erythrocyte with a water potential of −500 kPa was placed in solutions with a range of water potentials. The water potentials are listed below. Describe what would happen to the erythrocyte in each case.
a −470 kPa **b** −500 kPa **c** −1200 kPa *(3 marks)*

3 Suggest why some plants wilt when the soil they are growing in becomes too salty. *(4 marks)*

Key term

Osmosis: Diffusion of water through a partially permeable membrane down a water potential gradient.

Revision tip: Nothing higher than zero

Pure water (without any solute dissolved) has a water potential of 0. This is the highest possible value of ψ. Solutions always have a negative water potential. As more solute dissolves, water potential drops.

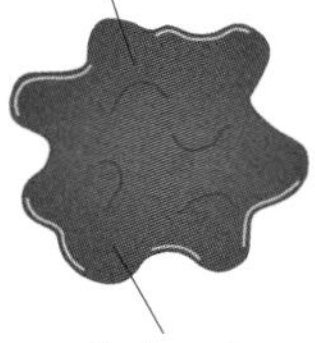

▲ **Figure 1** *Crenation (cell shrinkage) in an erythrocyte*

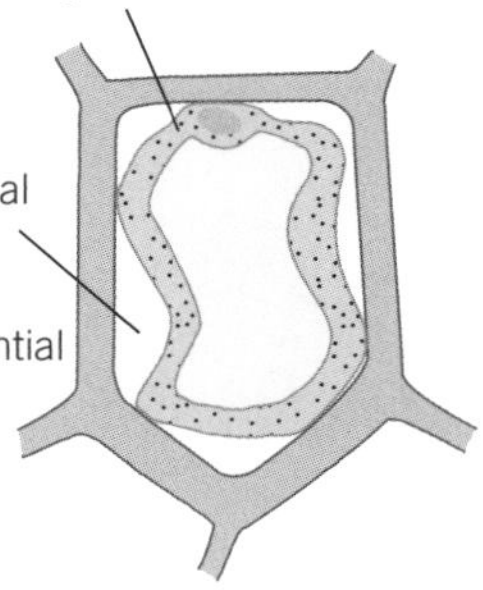

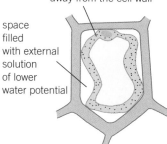

▲ **Figure 2** *Plasmolysis in a plant cell*

Chapter 5 Practice questions

1 Which of the following sentences is the most accurate description of the role of cholesterol within membranes?

A Cholesterol increases membrane fluidity

B Cholesterol limits the fluidity of a membrane at low temperatures

C Cholesterol regulates membrane fluidity

D Cholesterol prevents membrane fluidity from increasing *(1 mark)*

2 Which of the following substances is an example of a non-polar solvent?

A Benzene

B Water

C Cholesterol

D Phospholipid *(1 mark)*

3 Which of the following statements is true of facilitated diffusion?

A Particles move up a concentration gradient (from a low to a high concentration)

B Only occurs through channel proteins

C Does not require ATP

D Rate is independent of temperature *(1 mark)*

4 Which is the most appropriate term to describe the release of neurotransmitters from a neuronal cell?

A Endocytosis

B Exocytosis

C Pinocytosis

D Phagocytosis *(1 mark)*

5 Explain how the structure of membranes enables them to perform the following functions:

- Cell communication
- The regulation of the exchange of substances in and out of organelles. *(5 marks)*

6 Explain why membranes are described using a fluid mosaic model. *(2 marks)*

7 The following passage describes the movement of water via osmosis. Complete the passage by choosing the most appropriate word to place in each gap.

Water diffuses across a membrane from the side with the .... water potential to the side with the .... water potential. Plant cells that lose water will exhibit reduced .... and may undergo .... (i.e. the plasma membrane pulls away from the cell wall).

(4 marks)

8 Describe and explain the movement of water between the following three cells, which share boundaries with each other.

Cell A (water potential of −250 kPa)

Cell B (water potential of −180 kPa)

Cell C (water potential of −280 kPa) *(3 marks)*

6.1 The cell cycle

Specification reference: 2.1.6(a) and (b)

Having learned about the structure and components of cells in previous chapters, you will now explore how cells divide and differentiate. We begin in this topic by examining how a sequence of events, known as the cell cycle, enables a cell to divide.

Synoptic link

You learned about DNA replication in Topic 3.9, DNA replication and the genetic code.

The phases of the cell cycle

Before dividing, cells must grow and synthesise new organelles and molecules. These processes occur in a particular order, which is known as the cell cycle. Cells are in interphase, which is when growth and synthesis occurs, for the majority of the cycle. Mitosis (division of the cell nucleus) and cytokinesis (cell division) (see Topic 6.2, Mitosis) follow interphase. The key features of the cell cycle phases are illustrated in the following diagram.

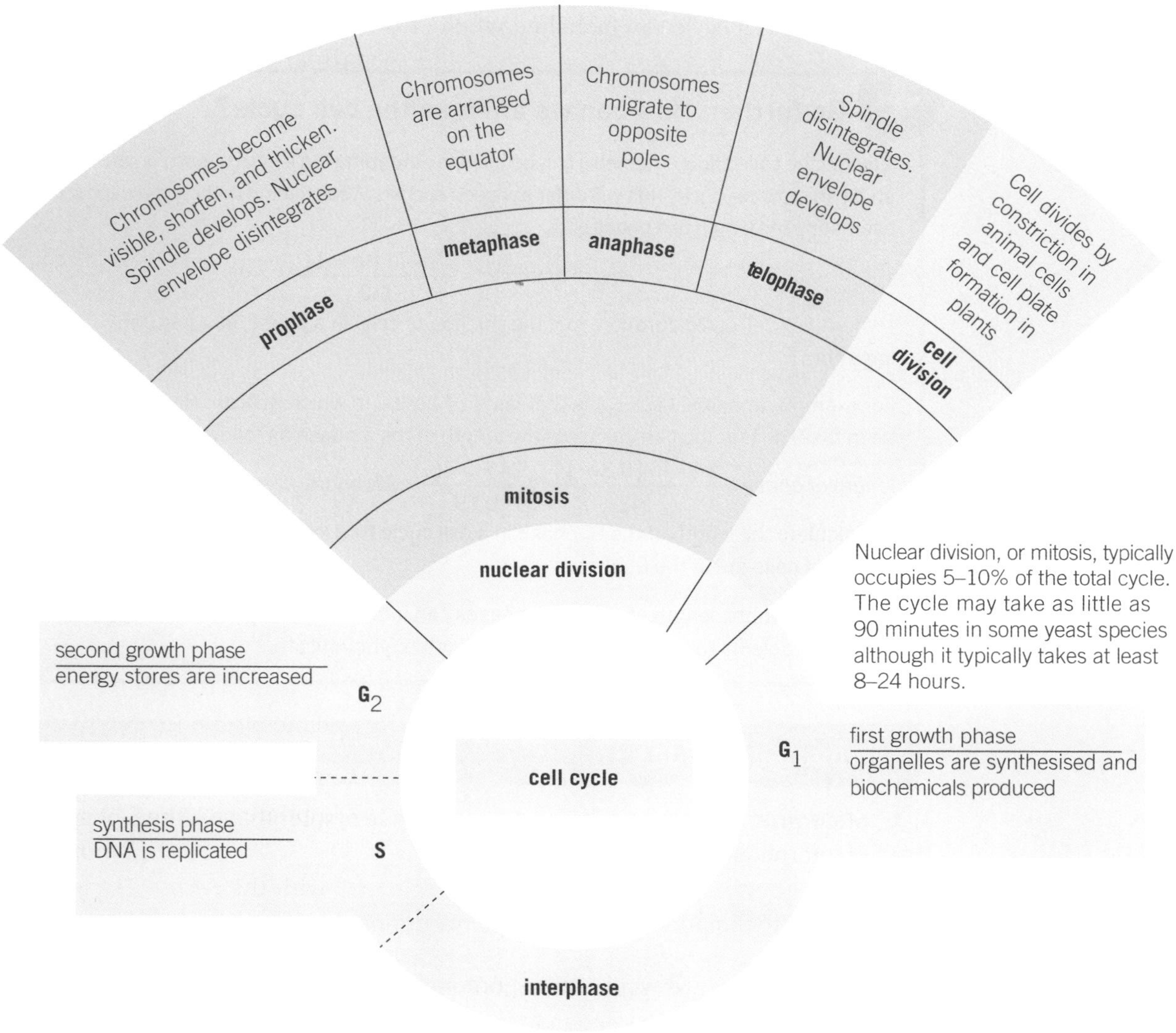

▲ **Figure 1** *Interphase comprises the majority of the cell cycle. The duration of one cycle varies between species and cell types*

Common misconception: Interphase is not a resting phase

Interphase is sometimes referred to as the 'resting phase'. This is misleading; in fact, it could not be further from the truth. A great deal of chemical activity, including the synthesis of organelles and DNA, takes place during interphase.

However, cells sometimes leave the cell cycle and stop dividing. For example, some mature, differentiated cells (see Topic 6.4, The organisation and specialisation of cells), such as neurons, no longer divide. The cycle can also be halted in cells with damaged DNA. These cells are said to have entered the G_0 phase. 'Resting phase' would be a more appropriate label for the G_0 phase rather than interphase.

Cell cycle control

The order and timing of processes in the cell cycle are under tight control. The cycle has checkpoints (e.g. at the end of the G_1 and G_2 phases) that verify whether each phase of the cycle has been completed correctly. These checkpoints are controlled by proteins called cyclins and cyclin-dependent kinases. The cell cycle can be halted when errors are detected at a checkpoint.

Go further: How can we analyse the cell cycle?

A technique called flow cytometry can be used to determine the length of each phase in an organism's cell cycle. This will differ between species. A calculation that can be used for estimating the length of a phase is:

$$\textit{length of phase} = \frac{T_c \times \ln(\text{fp} + 1)}{\ln 2}$$

[Where T_c = cell cycle duration, fp = the fraction of cells in a phase, ln = natural logarithm]

For example, imagine a cell cycle that lasts 24 hours, in which 30% of cells are found to be in the S phase. You can calculate the length of the S phase as follows:

$$\textit{length of phase} = \frac{24 \times \ln(0.3 + 1)}{\ln 2} = \frac{24 \times 0.26}{0.69} = 9.0 \text{ hours}$$

1 Calculate the length of the G_1 phase in a cell cycle that lasts 18 hours and in which 45% of cells are in the G_1 phase.

2 Changes in the length of cell cycle phases can indicate health problems. Suggest what problems these changes in cycle length might indicate.

Summary questions

1 State what is produced during **a** the G_1 phase of interphase **b** the S phase of interphase. *(3 marks)*

2 Explain the importance of checkpoints during the cell cycle. *(3 marks)*

3 **a** Suggest why it would be inappropriate to use the term 'resting phase' in relation to a cell in the cell cycle.
b Suggest which types of cells could be considered to be in a 'resting phase'. *(3 marks)*

6.2 Mitosis

Specification reference: 2.1.6 (c), (d), and (e)

You learnt in Topic 6.1, The cell cycle, that cells spend most of their time growing and synthesising molecules, in addition to performing specialised functions. The division of a cell takes up a relatively small proportion of its cycle, yet it is a crucial process. Here you will examine the steps involved in nuclear division (mitosis) and cell division (cytokinesis).

Key term

Mitosis: Nuclear division, producing two genetically identical cells.

Synoptic link

You learned about chromosome structure in Topic 2.4, Eukaryotic cell structure.

The importance of mitosis

Mitosis produces two nuclei that contain identical genetic material. This ensures that, following cell division, the two daughter cells are exact copies of the original cell. The replication of genetically identical cells is important for the following processes:

- *Growth*
- *Repair* (of damaged cells)
- *Replacement* (of cells, such as red blood cells, that have limited lifespans)
- *Asexual reproduction* (in eukaryotes).

The stages of mitosis

Mitosis is preceded by interphase. DNA replicates during interphase; therefore a cell at the start of mitosis has two copies of all its genetic material (i.e. two identical DNA molecules, called chromatids).

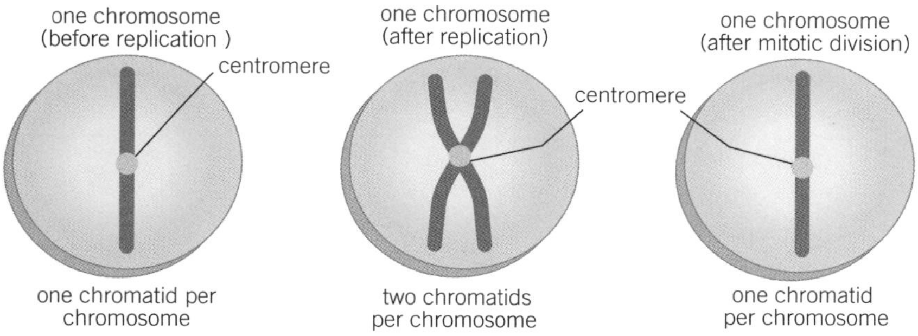

▲ **Figure 1** *Chromosome structure at different stages of the cell cycle*

Revision tip: Don't let confusion strand you

You should be careful when using the terms 'strand' and 'molecule'. A DNA molecule contains two polynucleotide strands. Each chromosome consists of two chromatids. Each chromatid is a double-stranded DNA molecule, not an individual strand.

Revision tip: Mitosis in plants

Mitosis is different in plants and animals.

Mitosis in animals	Mitosis in plants
Centrioles	No centrioles (in most plants)
Cells become rounded before division	No shape change
Spindle disappears before cytokinesis	Some of the spindle remains during cytokinesis

Stage	Key features	Appearance
Prophase	• Chromatin (comprising DNA and histone proteins) condenses – **chromosomes become visible**. • **Nucleolus** disappears. • **Centrioles** move to the poles of the cell. • **Nuclear envelope** breaks down (towards the end of prophase).	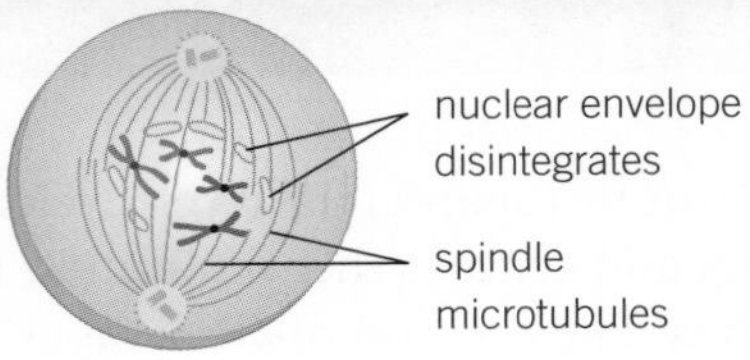
Metaphase	• **Spindle fibres** (organised by the centrioles) attach to **centromeres** (towards the end of prophase/beginning of metaphase). • **Chromosomes line up** along the centre (equator) of the cell.	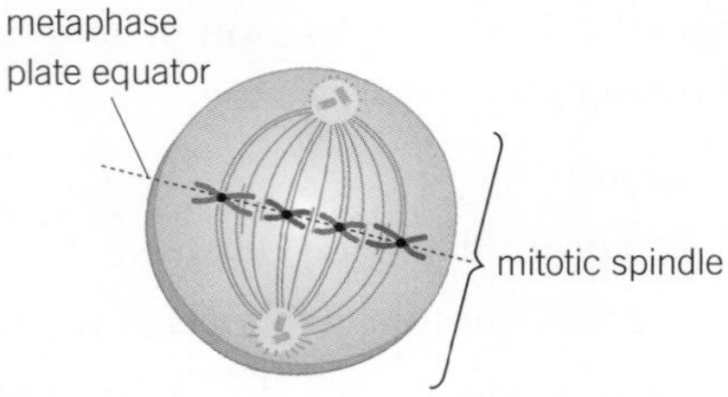
Anaphase	• Spindle fibres shorten. • **Centromeres divide**. • **Chromatids are separated** and pulled to opposite poles of the cell.	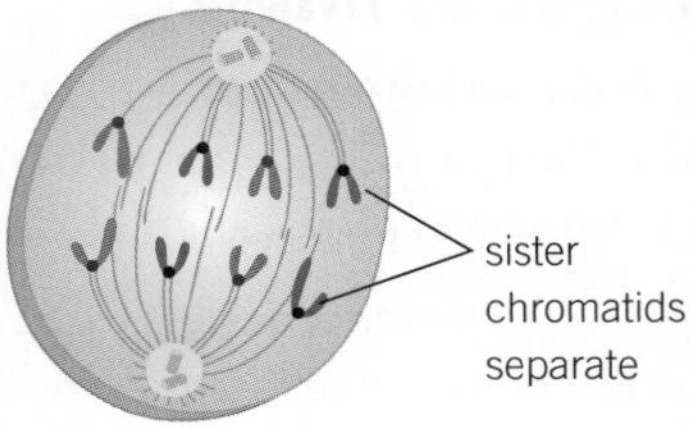
Telophase	• Chromatids are at the poles of the cell (and can be referred to as daughter chromosomes). • **Nuclear envelopes reform** around each set of chromosomes. • **Chromosomes uncoil** (and are no longer visible). • Cell division (**cytokinesis**) begins.	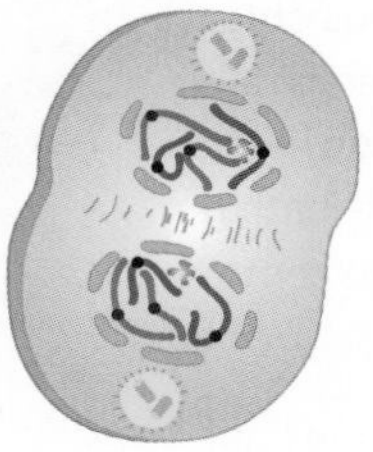

Cytokinesis

The division of a cell (cytokinesis) begins in telophase. This results in two genetically identical cells. Each cell receives approximately half of the organelles and cytoplasm from the original cell.

Different methods of cytokinesis operate in animal and plant cells. This is principally a consequence of plants having cell walls.

In **animals:** a *cleavage furrow* forms (i.e. cell surface membranes are pulled together by the *cytoskeleton*).

In **plants:** the cell wall prevents cleavage furrows. The two daughter cells are instead separated by new *cell wall production* down the centre of the original cell.

Summary questions

1 Human somatic cells have a diploid number of 46. State the number of chromatids present in a human cell **a** during prophase **b** at the end of cytokinesis. *(2 marks)*

2 Plants lack centrioles. What does this indicate about the role of centrioles in mitosis? *(2 marks)*

3 The calculation of mitotic index provides an indication of the rate of cell division in a tissue. *Mitotic index* = $\frac{\textit{number of cells in mitosis}}{\textit{total number of cells}}$ Suggest which disease might result in an increase in mitotic index. Explain your answer. *(2 marks)*

6.3 Meiosis

Specification reference: 2.1.6(f) and (g)

Mitosis, as you learned in Topic 6.2, Mitosis, is nuclear division that produces two identical nuclei. The formation of reproductive cells, however, creates genetic variation and involves a different type of nuclear division called meiosis.

The stages of meiosis

Meiosis I separates homologous chromosomes to produce two haploid cells. Meiosis II is similar to mitosis: chromatids are separated to form a total of four haploid daughter cells.

Stage	Key features	Appearance
Prophase I	The key events of mitotic prophase (see Topic 6.2, Mitosis) occur (i.e. nuclear envelope disintegrates, nucleolus disappears, spindles form, chromosomes condense). Homologous chromosomes pair up (form **bivalents**). **Crossing over**	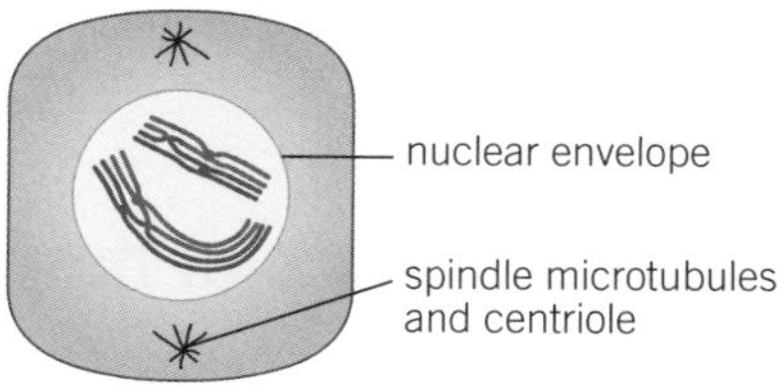
Metaphase I	Homologous pairs (bivalents) line up at the cell equator. **Independent assortment** of chromosomes.	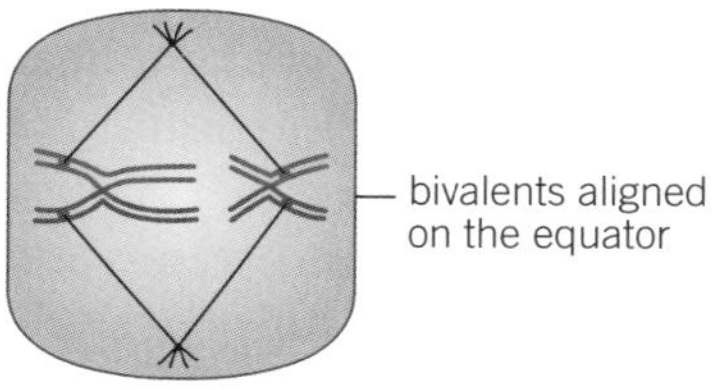
Anaphase I	Homologous chromosome pairs are separated (**random segregation**) – sister chromatids both remain attached to centromeres.	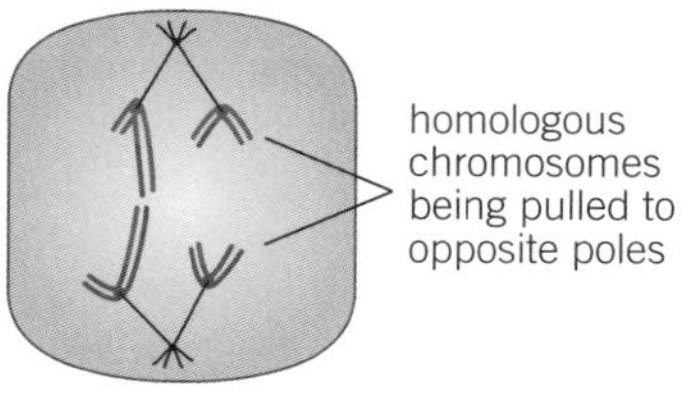
Telophase I	Chromosomes assemble at either pole. **Cytokinesis** (cell division) – two haploid cells formed.	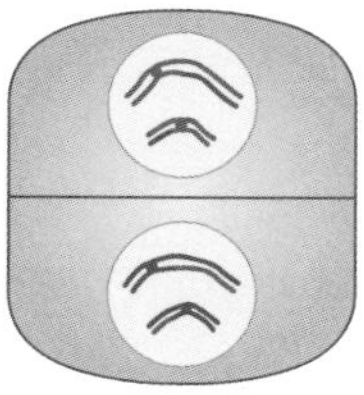
Prophase II	Chromosomes condense, nuclear envelope breaks down, and spindles form.	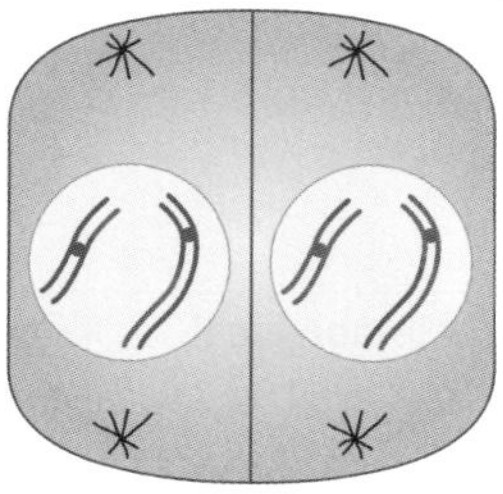
Metaphase II	Chromosomes line up at equator (as in mitosis). **Independent assortment** of chromatids.	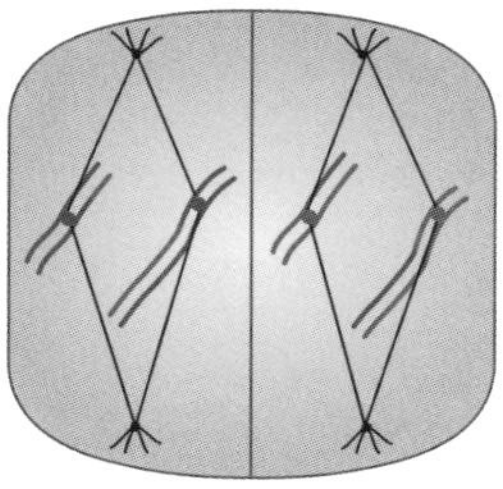

Revision tip: Homologous chromosomes

Each nucleus contains pairs of matching chromosomes, which have the same gene loci (but possibly different versions of those genes, i.e. different alleles). These pairs are known as homologous chromosomes. Each pair comprises a maternal chromosome and a paternal chromosome. Homologous pairs are the same length and, at the start of nuclear division, each chromosome consists of two chromatids.

Key term

Meiosis: A form of nuclear division in which the chromosome number is halved.

Synoptic link

We explored DNA structure in Topic 3.9, DNA replication and the genetic code.

Anaphase II	**Random segregation** of chromatids.	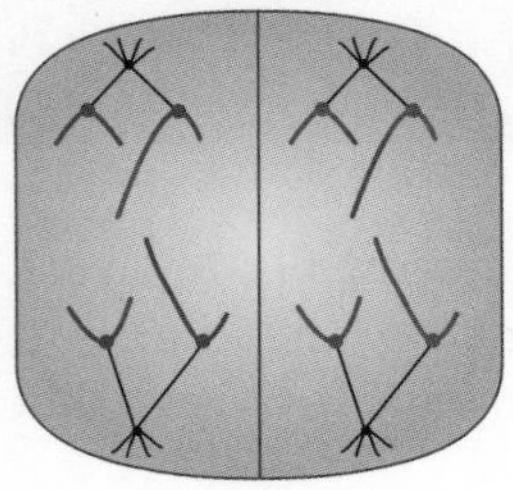
Telophase II	Chromatids assemble at poles. **Cytokinesis** (cell division) – four haploid daughter cells formed.	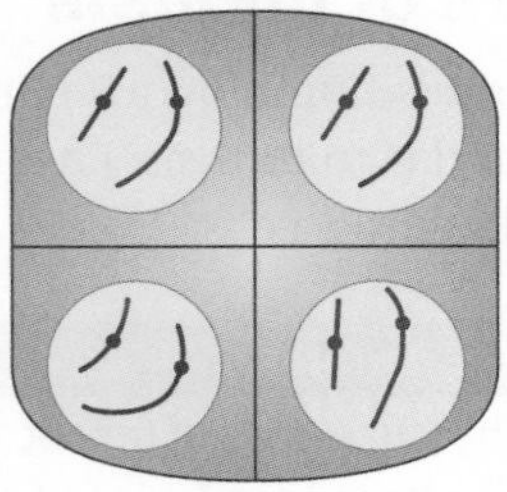

The importance of meiosis

Species that undergo asexual reproduction produce genetically identical offspring. Many multicellular organisms, however, reproduce sexually. These species produce sex cells (gametes) that contain half the standard number of chromosomes (i.e. the haploid number). Meiosis, which is known as a reduction division, enables the chromosome number to be halved. As a result, the standard chromosome number (i.e. the diploid number) is restored in the offspring when two organisms reproduce.

As well as ensuring the preservation of chromosome number through generations, meiosis introduces genetic variation. This variation is introduced through the processes outlined in the following table:

Process	Stage of meiosis	What happens?	How does this produce variation?
Crossing over	Prophase I	Non-sister chromatids (chromatids from different chromosomes in a homologous pair) interweave (at points called chiasmata), forming bivalents.	Genetic material is exchanged between homologous chromosomes, producing new combinations of alleles.
Independent assortment of chromosomes	Metaphase I	When homologous chromosomes move to the cell equator, the alignment of each chromosome (i.e. on which side of the cell it is positioned) is random.	When homologous chromosomes are separated in anaphase I, many different chromosome combinations can be formed in daughter cells.
Independent assortment of chromatids	Metaphase II	Chromosomes line up at the cell equator. The side on which each sister chromatid is positioned is random.	Sister chromatids are no longer identical (due to crossing over), therefore many chromatid combinations are possible in daughter cells.

Revision tip: Random and independent?

Independent assortment (during metaphase) can also be called random assortment. Similarly, random segregation (during anaphase) can be called independent segregation.

Revision tip: Sister or not?

Sister chromatids are joined by a centromere on the same chromosome and are identical to each other prior to crossing over. Non-sister chromatids are on different chromosomes in a homologous pair – they have the same gene loci but are not identical (i.e. they have different alleles). Non-sister chromatids cross over during prophase I.

Summary questions

1 Describe the differences between **a** mitosis and meiosis II **b** meiosis I and meiosis II. *(4 marks)*

2 The number of possible combinations of chromosomes in a gamete can be calculated using the term 2^n, with 'n' being the haploid number of the species. Tigers (*Panthera tigris*) have a diploid chromosome number of 38. Calculate the number of different chromosome combinations that are possible in a tiger gamete. *(1 mark)*

3 Explain how genetic variation is introduced to gametes during meiosis. *(4 marks)*

6.4 The organisation and specialisation of cells

Specification reference: 2.1.6(h) and (i)

Although they share common features, cells in an organism are not identical; different cells become specialised for particular roles, depending on their positions within the organism.

Specialised cells

In order to specialise for a particular function, cells differentiate. All cells in an organism contain the same DNA; differentiation requires a cell to express certain genes but switch off others. The table summarises the adaptations exhibited by some specialised cells.

	Structural feature	Function
Red blood cells/ erythrocytes	Flattened, biconcave shape. No organelles. Flexible due to protein arrangements in membranes.	Increases the surface area: volume ratio to increase the rate of O_2 diffusion. More space available for haemoglobin. Ability to squeeze through capillaries.
Neutrophils (a type of white blood cell)	Multi-lobed nucleus. Many lysosomes.	Ability to squeeze through gaps in capillary walls to reach infections. They contain hydrolytic enzymes (to destroy pathogens).
Sperm cells	Flagellum (tail). Many mitochondria. Acrosome. 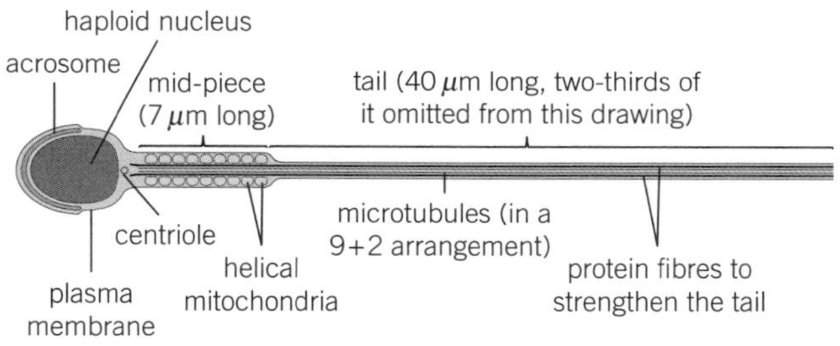	For movement towards the egg. To supply energy for movement. Contains digestive enzymes to enable the sperm to penetrate the egg.
Palisade cells	Many chloroplasts (packed close together). Thin cell walls. Large vacuole.	High rate of light absorption for photosynthesis. Greater CO_2 diffusion rate. To maintain turgor pressure.
Root hair cells	Long, narrow extensions of the cell. Large vacuole with a high concentration of dissolved solutes.	Large surface area to increase water and mineral uptake from the soil. Lower water potential to increase rate of water uptake from the soil.
Guard cells	Two kidney-shaped cells with thickened inner cell walls 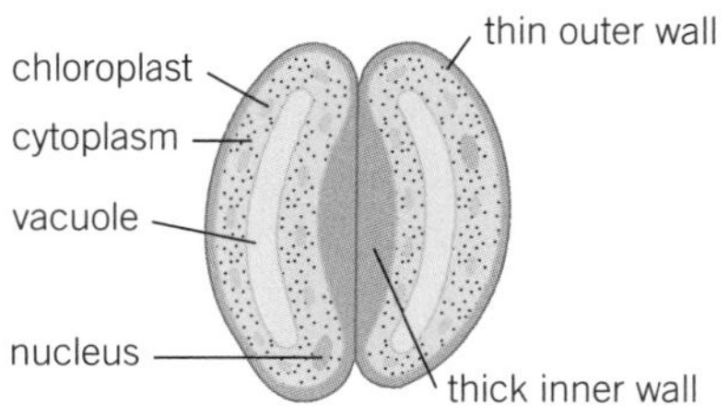	They control when stomata open (depending on the requirement for gas exchange and water levels in the plant)

Synoptic link

You will learn more about xylem and phloem tissue in Chapter 9, Transport in plants.

Tissues

A tissue is a collection of differentiated cells that work together for a particular function. Examples include:

Tissue	Location	Structural features	Function
Squamous epithelium	Capillaries and alveoli	Thin/flat	Increases diffusion rate
Ciliated epithelium	Trachea	Cilia on the outside of cells	Sweep mucus from trachea
Cartilage	Joints (between bones)	Firm and flexible	Protective connective tissue
Skeletal muscle	Attached to bones	Contractile proteins	Movement of the skeleton
Xylem	Plant stem	Elongated dead cells strengthened by lignin	Transport of water
Phloem	Plant stem	Perforated walls	Transport of nutrients

Organs

Organs are collections of several tissues that combine to perform a function or range of functions. For example, the heart comprises squamous epithelium, endothelium, and cardiac muscle, as well as other tissues.

Organ systems

Organs that work in conjunction with each other can be considered an organ system. For example, the digestive system includes the following organs: oesophagus, stomach, liver, pancreas, gall bladder, gastrointestinal tract.

Overall, an organisational hierarchy exists:

Specialised cells – tissues – organs – organ systems – whole organism.

Summary questions

1 Outline how a sperm cell is specialised for its function. *(4 marks)*

2 Explain how the presence of squamous epithelium tissue improves the efficiency of gas exchange in the lungs. *(4 marks)*

3 Sclerenchyma tissue provides support to plants. Suggest some features of sclerenchyma tissue that enable it to carry out its function. *(3 marks)*

6.5 Stem cells

Specification reference: : 2.1.6(j), (k), (l), and (m)

You looked at examples of differentiated cells in Topic 6.4, The organisation and specialisation of cells. Here we discuss the undifferentiated cells from which specialised cells arise. These undifferentiated cells are known as stem cells.

What are stem cells?

Although they are undifferentiated, stem cells vary in the range of cells they have the potential to form; this is referred to as their potency. The following table outlines the traits of different types of stem cells.

Potency of stem cell	Which cells can they divide to form?	Examples
Totipotent	Any (and they have the potential to form a whole organism)	The first 16 cells of an animal zygote. Plant meristem cells (including cambium tissue, which differentiates into xylem and phloem tissue).
Pluripotent	All tissues (but not a whole organism)	Early embryonic cells (in the blastocyst)
Multipotent	A limited range of cells	Haematopoietic stem cells (in bone marrow), which can differentiate to form all blood cells, including erythrocytes and neutrophils

Uses of stem cells

Current uses include:

- Drug testing *in vitro* (i.e. testing drugs on cultured cells in a laboratory)
- Studying developmental biology and disease development *in vitro*
- Treatment of burns
- Bone marrow transplants to replace stem cells destroyed during cancer treatment.

Future diseases that could be treated include: Parkinson's and Alzheimer's disease (using neural stem cells), heart disease, and type 1 diabetes.

Ethics

Some people have moral and religious reservations about the use of pluripotent stem cells obtained from embryos. Scientists are developing techniques for reprogramming differentiated adult cells back into pluripotent stem cells (known as induced pluripotent stem cells (iPSCs)). These techniques do not require the use of embryos and would therefore overcome the ethical objections held by some people.

Summary questions

1 Describe one similarity and one difference between totipotent and pluripotent stem cells. *(2 marks)*

2 Explain the potential benefits of using induced pluripotent stem cells rather than pluripotent embryonic stem cells for disease treatment in the future. *(3 marks)*

3 Suggest why plants are more able than animals to form reproductive clones. *(3 marks)*

Chapter 6 Practice questions

1 Crossing over during meiosis contributes to genetic variation.

Which is the correct description of crossing over?

A Chiasmata form when chromatids on non-homologous chromosomes exchange genetic material in prophase I.

B Chiasmata form when chromatids on homologous chromosomes exchange genetic material in prophase I.

C Chiasmata form when chromatids on non-homologous chromosomes exchange genetic material in metaphase I.

D Chiasmata form when chromatids on homologous chromosomes exchange genetic material in metaphase I. *(1 mark)*

2 Mesenchymal stem cells can differentiate into adipose, bone, cartilage, and connective tissue cells, but not other types of cell.

What type of stem cells are mesenchymal stem cells?

A Unipotent **C** Pluripotent

B Multipotent **D** Totipotent *(1 mark)*

3 One method for estimating the length of a cell cycle phase is:

$$\text{Length of phase} = \frac{[T_C \times \ln(fp + 1)]}{\ln 2}$$

(where T_C = cell cycle duration, fp = the fraction of cells in a phase, ln = natural logarithm)

What is the length of the G_2 phase in a cell cycle that lasts 16 hours and in which 35% of cells are in the G_2 phase?

A 3 hours **C** 7 hours

B 6 hours **D** 8 hours *(1 mark)*

4 Describe the potential uses of stem cells in disease treatment and the problems that have been encountered when researching potential stem cell therapies. *(6 marks)*

5 The following passage describes prophase I of mitosis. Complete the passage by choosing the most appropriate word to place in each gap.

.... (i.e. DNA and its associated proteins) condenses in prophase I. The .... and nuclear membrane disappear. .... move to opposite poles of the cell. These structures are composed of ....

(4 marks)

6 The letters A–D list features of specialised cells. Match each of the features to the specialised cell (W–Z) being described. In each case, explain the benefit of the adaptation.

A Thin cell walls **W** Neutrophil

B Acrosome **X** Erythrocyte

C No organelles **Y** Sperm cell

D Many lysosomes **Z** Palisade cell *(8 marks)*

7.1 Specialised exchange surfaces

Specification reference: 3.1.1(a) and (b)

Single-celled organisms obtain oxygen through diffusion, which you learnt about in Topic 5.3, Diffusion. Larger, multicellular organisms, however, cannot rely on diffusion alone; they have evolved specialised gas exchange surfaces.

The need for specialised exchange surfaces

Multicellular organisms have evolved exchange surfaces because:

- Metabolic activity is greater in multicellular organisms, which means oxygen needs to be supplied and carbon dioxide removed at higher rates.
- Surface area to volume (SA:V) ratios are smaller in multicellular organisms (i.e. volume increases in relation to surface area), which means that diffusion alone would not achieve an adequate rate of gas exchange.

Synoptic link

You learned about the passive movement of substances in Topic 5.3, Diffusion.

Features of exchange surfaces

We will explore examples of exchange surfaces that have evolved in different species in Topic 7.2, The mammalian gaseous exchange system, and Topic 7.4, Ventilation and gas exchange in other organisms. All of these adaptations have features in common.

▼ **Table 1** *Common features of gas exchange surfaces*

Feature of exchange surfaces	Benefit
Increased surface area	Overcomes the reduced SA:V ratio in larger organisms
Thin layers	Reduces the diffusion distance
Good blood supply	Maintains steep concentration gradients through the quick supply and removal of gases
Good ventilation	

Maths skill: SA:V ratios

Surface area to volume ratio decreases as organisms increase in size. We can demonstrate this effect by tracking the changes in surface area and volume of spheres as they increase in size.

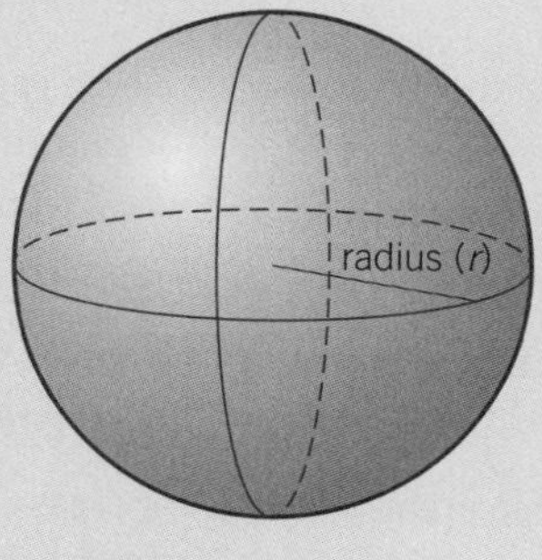

Surface area of a sphere = $4\pi r^2$

Volume of a sphere = $\frac{4}{3}\pi r^3$

For example, a sphere with a radius of 1 m would have a surface area of 12.6 m^2 (4 × 3.14 × 1 × 1) and a volume of 4.2 m^3 (4/3 × 3.14 × 1 × 1 × 1). When the radius is increased to 2 m, the proportional difference between the surface area (50.2 m^2) and volume (33.5 m^3) decreases. The SA:V ratio becomes smaller.

Summary questions

1 Explain why multicellular organisms require specialised gas exchange surfaces. *(4 marks)*

2 Describe how specialised gas exchange surfaces maintain concentration gradients to enable a high rate of diffusion. *(4 marks)*

3 Actinophryids are spherical microorganisms. A population of one genus of actinophryids (*Actinophrys*) has a mean diameter of 45 μm. Another genus (*Actinosphaerium*) has a mean diameter of 400 μm. Calculate the SA:V ratios for the two genera. *(6 marks)*

7.2 The mammalian gaseous exchange system
7.3 Measuring the process

Specification reference: 3.1.1(c), (d), and (e)

Revision tip: How do surfactants work?
Water is present on the inner surface of alveoli. Oxygen dissolves in the water before diffusing across the alveolar epithelium layer. If only water were present, however, the alveoli would collapse. A surfactant (a mixture of phospholipids and protein) prevents alveoli collapsing by interfering with hydrogen bond formation between water molecules.

Mammals have evolved a gaseous exchange system that performs a balancing act: limiting the amount of water lost from the body while enabling efficient gas exchange. The exchange of gases in mammals occurs in the lungs, the features of which are our focus in this topic.

Structures in the mammalian gas exchange system

The mammalian gas exchange system consists of passages (the nasal cavity, trachea, bronchi, and bronchioles) and the gas exchange surfaces (alveoli in the lungs) to which they deliver air. The principal features of these structures are outlined in the following table.

Structure	Key features	Function
Nasal cavity	Good blood supply	Warms air entering the body
	Lined with hairs and **mucus**-secreting cells	Traps dust and bacteria (protection from disease)
	Moist surface	Increases humidity, reducing evaporation from the lungs
Trachea	Supported by flexible **cartilage**	Prevents collapse
	Lined with **goblet cells**, which secrete mucus	Trap dust and bacteria
	Ciliated epithelium cells	Cilia move mucus away from the lungs
Bronchus	Cartilage, like the trachea	Prevents collapse
Bronchioles	**Smooth muscle** (and no cartilage)	Bronchioles can constrict and dilate to vary the amount of air reaching the lungs
	Flattened epithelium cells	Some gaseous exchange is possible
Alveoli	Single layer of **flattened epithelium** cells	Short diffusion pathway, which increases diffusion rate
	Elastic fibres and **collagen**	Enable stretching and elastic recoil during ventilation
	Large surface area	Increased rate of diffusion (*see Topic 5.3, Diffusion, and Topic 7.1, Specialised exchange surfaces*)
	Good blood supply (alveoli are surrounded by a capillary network) and good ventilation	O_2 is supplied to the alveoli and moved into the circulatory system quickly. CO_2 is supplied from the circulatory system and removed from the lungs quickly. This maintains a **steep concentration gradient**.
	Covered with a layer of **surfactant**	Alveoli remain inflated

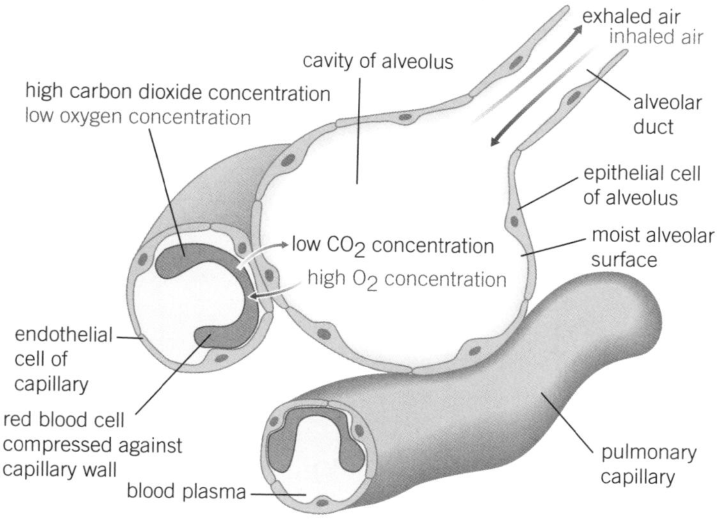

▲ **Figure 1** *Alveolar gas exchange*

Mechanism of ventilation

Air moves in and out of the lungs because of changes in pressure; this is known as ventilation. Efficient ventilation (i.e. the efficient supply of oxygen to the alveoli and the quick removal of carbon dioxide) maintains a steep concentration gradient in the lungs. The following table compares the changes that occur in the lungs during inspiration and expiration.

Inspiration (inhalation)	Expiration (exhalation)
External intercostal muscles contract	External intercostal muscles relax
Ribs move up and out	Ribs move down and inwards
Diaphragm contracts and flattens	Diaphragm relaxes and reverts to its domed shape
Thorax volume increases	Thorax volume decreases
Air pressure in the lungs drops below atmospheric pressure	Air pressure in the lungs rises above atmospheric pressure
Air moves into the lungs	Air moves out of the lungs

Common misconception: Relax, you're exhaling normally ...

Inspiration is an active process (i.e. external intercostal muscles and the diaphragm contract). Normal expiration (breathing out when at rest) is passive; the diaphragm and external intercostal muscles relax and return to their resting positions. Forceful exhalation (e.g. during physical exertion or coughing), however, requires muscular contraction to move air from the lungs; the internal intercostal muscles contract, pulling the ribs down, and abdominal muscles force the diaphragm back to its domed position.

How is lung function measured?

Measurements of lung function tend to be made using two pieces of equipment: a peak flow meter and a spirometer.

Equipment	Method	Use
Peak flow meter	Measures the rate at which patients expel air into a handheld tube.	Can be used to monitor conditions such as asthma.
Spirometer	Patients breathe in and out of a mouthpiece attached to a sealed chamber; oxygen from the chamber is used up.	Can measure several aspects of lung volume (see below).

Revision tip: Defining vital capacity

Vital capacity can potentially be defined and measured in two ways. It is either the maximum volume *exhaled* following the strongest possible *inhalation* or the maximum volume *inhaled* following the strongest possible *exhalation*. The values should be the same.

Worked example: Breathing calculations

Breathing rate = the number of breaths taken per minute.

Ventilation rate (the total volume inhaled per minute) = breathing rate × tidal volume

You may be asked to rearrange this equation. For example, what is the tidal volume of a person with a ventilation rate of 6.5 $dm^3 min^{-1}$ and a breathing rate of 13 breaths min^{-1}?

Tidal volume = ventilation rate / breathing rate = 6.5 / 13 = 0.5 dm^3 (500 cm^3).

What is measured in spirometry?

A variety of measurements can be made using a spirometer.

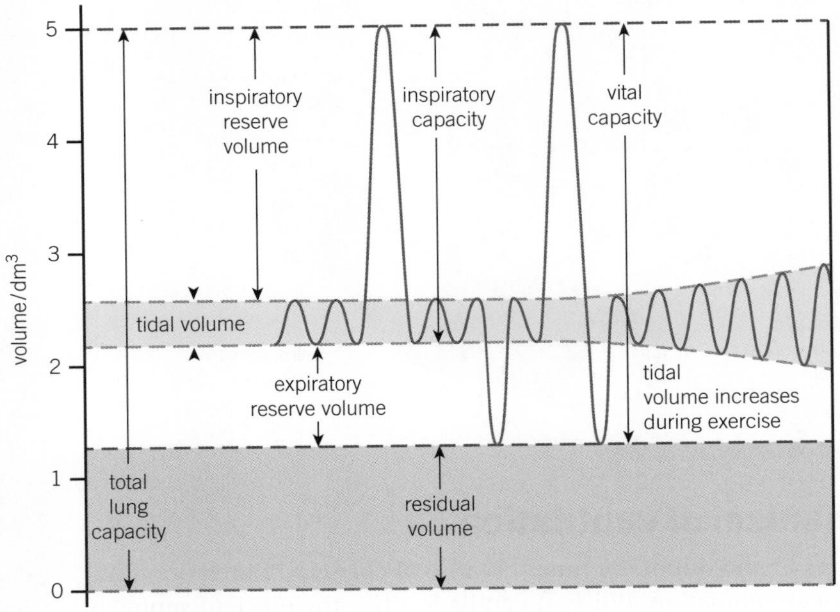

▲ **Figure 2** *Lung capacity has many measurable components*

- **Total lung capacity** = vital capacity + residual volume.
- **Residual volume** is the volume remaining in the lungs even after a person has exhaled with maximum force.
- **Vital capacity** is the maximum volume that can be breathed out following the strongest possible inhalation (i.e. tidal volume + inspiratory reserve volume + expiratory reserve volume).
- **Tidal volume** is the volume inhaled with each resting breath (or the volume exhaled with each resting breath).
- Inspiratory reserve and expiratory **reserve volumes** are the additional volumes of air that can be breathed in and out during forced inhalation and exhalation.

Summary questions

1 State three features of alveoli that enable a high rate of gas exchange. *(3 marks)*

2 Calculate the breathing rate of a person with a ventilation rate of 6.20 $dm^3 min^{-1}$ and a tidal volume of 0.52 dm^3. Give your answer to two significant figures. *(2 marks)*

3 The partial pressure (kPa) of a gas reflects its relative concentration in a gas mixture. The table below shows partial pressures of oxygen and carbon dioxide in three parts of the mammalian body.

Gas	Alveolar air (kPa)	Blood in pulmonary artery (kPa)	Blood in pulmonary vein (kPa)
Oxygen	13.8	5.3	13.8
Carbon dioxide	5.3	6.1	5.3

a Explain why blood in the pulmonary vein has the same partial pressures as the air in the alveoli. *(2 marks)*

b Suggest an explanation for the differences in partial pressure of oxygen and carbon dioxide in the pulmonary artery and alveolar air. *(2 marks)*

7.4 Ventilation and gas exchange in other organisms

Specification reference: 3.1.1(f), (g), and (h)

In Topic 7.2, The mammalian gaseous exchange system, you read about the lungs, which are the organs of gas exchange in mammals. Different internal gas exchange systems have evolved in other groups of multicellular organisms. Here we discuss two of these systems: the tracheal system in insects and gills in bony fish.

Gaseous exchange in insects and fish

Insects evolved hard exoskeletons for structural support and to limit water loss. These exoskeletons prevent the diffusion of gas across insects' body surfaces. Insects also lack blood pigments capable of transporting oxygen. As a consequence, insects evolved a system of tubes (**tracheae**) that deliver oxygen directly to cells (and remove carbon dioxide).

Aquatic organisms have another difficulty to overcome. Water is considerably denser than air; moving water in and out of a lung-like structure would require too much energy. Instead, bony fish have evolved structures called **gills**, which extract oxygen from water flowing across them in one direction.

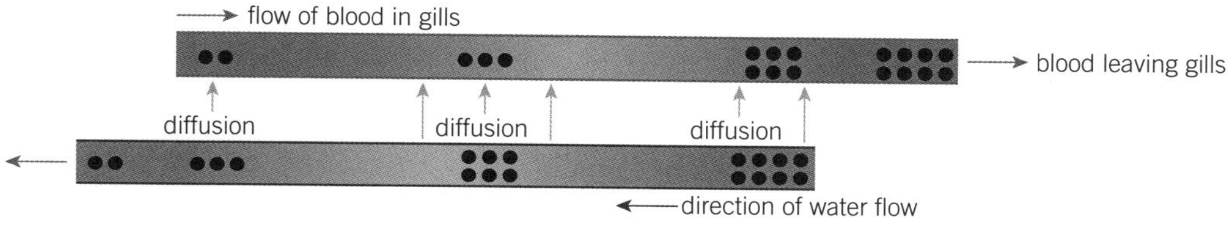

▲ **Figure 1** *Countercurrent flow in the gills. An oxygen concentration gradient is maintained. (Circles represent arbitrary units of oxygen concentration)*

Revision tip: Examining gas exchange systems

You may have the opportunity to dissect and examine the structures discussed here. You should be able to interpret diagrams and photomicrographs of the gill and tracheal tissue. Revising microscopy skills (Topic 2.1, Microscopy, and Topic 2.3, More microscopy) and magnification calculations (Topic 2.2, Magnification and calibration) will help.

Revision tip: Tracheae

You will have encountered tracheae in the context of mammalian gas exchange. *Trachea* means 'air pipe'; it is a tube that carries gases in and out of organisms. Mammals have one (singular: trachea), whereas insects have many (plural: tracheae).

▼ **Table 1** *A comparison of insect and fish gas exchange systems*

	Insect tracheae	Gills in bony fish
How does air/ water enter and leave the organism?	Air enters and leaves through small openings called **spiracles**. Spiracles are closed when possible to minimise water loss.	Water enters through the mouth. A continuous flow of water across the gills is achieved by the floor of the buccal cavity (mouth) being lowered (which takes water into the mouth) then raised (which forces water over the gills). Water leaves a fish when the operculum (a flap covering the gills) opens.
How is oxygen taken up?	Air travels through tracheae, which branch into tracheoles. Oxygen diffuses into cells from the tracheoles. Some active insects have high energy demands; they have evolved a muscular pumping system to increase oxygen supply.	Oxygen diffuses into gill **lamellae**, which are thin plates, packed with blood capillaries. Lamellae are attached to gill **filaments**. Water and blood (in lamellae) flow in opposite directions (i.e. **countercurrent flow**). This ensures an oxygen concentration gradient is maintained along the gill.
How does carbon dioxide leave?	CO_2 diffuses from tissues into the tracheae down a concentration gradient.	CO_2 diffuses from capillaries, across lamellae, and into water, which leaves through the operculum.
Appearance of the gas exchange system	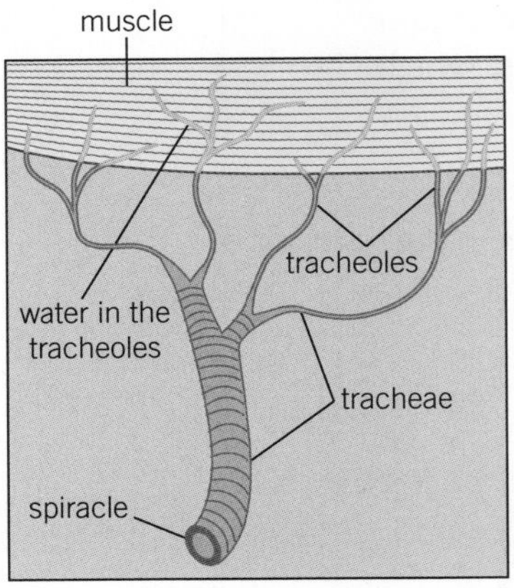	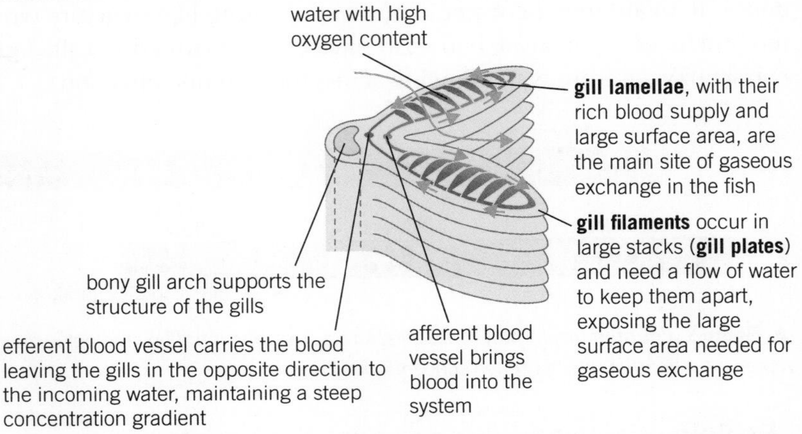

Summary questions

1. Explain why the evolution of a specialised gas exchange system was necessary for insects. *(2 marks)*
2. Compare the adaptations of alveoli and gills for gas exchange. *(6 marks)*
3. Explain how the structure of gills enables concentration gradients to be maintained during gas exchange. *(3 marks)*

Chapter 7 Practice questions

1 Which of the following is a prominent feature of the bronchioles?

A Smooth muscle

B Cartilage

C Goblet cells

D Ciliated epithelium *(1 mark)*

2 Which of the following is the correct description of tidal volume?

A The maximum volume of air that can be inhaled in addition to resting inhalation.

B The maximum volume of air that can be exhaled in addition to resting exhalation.

C The volume of air inhaled with each resting breath.

D The volume of air that remains in the lungs after forced exhalation.

(1 mark)

3 The breathing rate of a person is measured as 15 breaths per minute. The person's ventilation rate is 6.4 dm^3 per minute. Which of the following values represents the person's tidal volume?

A 2 cm^3

B 96 cm^3

C 427 cm^3

D 96 000 cm^3 *(1 mark)*

4 Air passes through several structures during gas exchange in insects. Which of the following sets of words represent the order in which air passes through the insect exchange system?

A Spiracle, tracheae, tracheoles

B Spiracle, tracheoles, tracheae

C Tracheae, tracheoles, spiracle

D Tracheoles, tracheae, spiracle *(1 mark)*

5 **a** A cell of the bacterial species *Streptococcus pyogenes* was analysed. The diameter of the cell was found to be 1 μm. Calculate the surface area to volume ratio of the cell. *(4 marks)*

b Explain why *Streptococcus pyogenes* does not require a specialised gas exchange system. *(2 marks)*

6 The following passage describes the inhalation of air in mammalian lungs. Complete the passage by choosing the most appropriate word to place in each gap.

The .... intercostal muscles contract; this causes the ribs to move up and out. The diaphragm contracts and .... The volume of the .... increases, and air .... in the lungs decreases below that of the atmosphere. *(4 marks)*

8.1 Transport systems in multicellular animals

Specification reference: 3.1.2(a) and (b)

Synoptic link

You considered the concept of surface area to volume ratio in Topic 7.1, Specialised exchange surfaces.

Revision tip: Gas transport in insects

The tracheal system (see Topic 7.4, Ventilation and gas exchange in other organisms) is the principal site of oxygen and carbon dioxide exchange in insects. However, respiratory gases can be carried, to a small extent, in the haemolymph of some species with open circulatory systems.

Due to their large size, multicellular animals have evolved exchange surfaces, which you learned about in Chapter 7, and internal transport systems, which distribute nutrients and oxygen around their bodies. Here we discuss the different transport systems that have evolved in animals.

Why are transport systems needed?

The rate at which substances can travel through multicellular animals by diffusion alone is too slow to meet their requirements. The evolution of internal transport systems enables substances to be circulated at a faster rate. Specialised transport systems are necessary because multicellular animals have:

- high metabolic rates (and high demand for oxygen and nutrients)
- relatively small surface area to volume ratios
- molecules (e.g. enzymes and hormones) that need to be transported to specific tissues.

Transport systems in animals

Internal transport (circulatory) systems can be labelled as either open or closed.

▼ **Table 1** *A comparison of open and closed circulatory systems*

	Open circulatory system	Closed circulatory system
Description	Few vessels; the transport medium (haemolymph) is pumped from the heart into the body cavity (haemocoel)	Transport medium (blood) is enclosed in vessels
Examples	Some invertebrates, such as arthropods (including insects) and molluscs	All vertebrates, and many invertebrates (e.g. cephalopods and annelid worms)

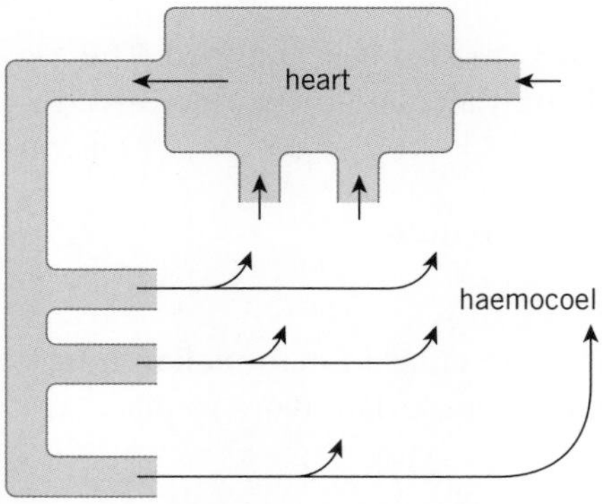

▲ **Figure 1** *A generalised open circulatory system*

Closed circulatory systems vary in their efficiency. Some species have a single circulatory system; others have evolved a double system, which enables a faster blood flow to be maintained. Species with double circulatory systems are able to sustain higher metabolic rates.

▼ **Table 2** *A comparison of single and double circulatory systems*

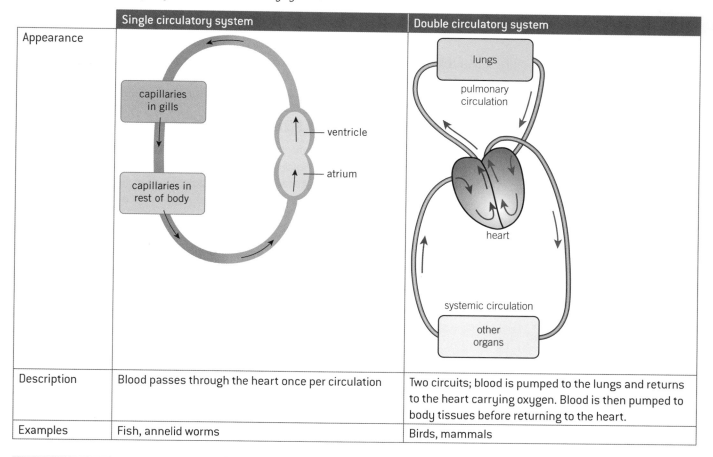

	Single circulatory system	Double circulatory system
Appearance	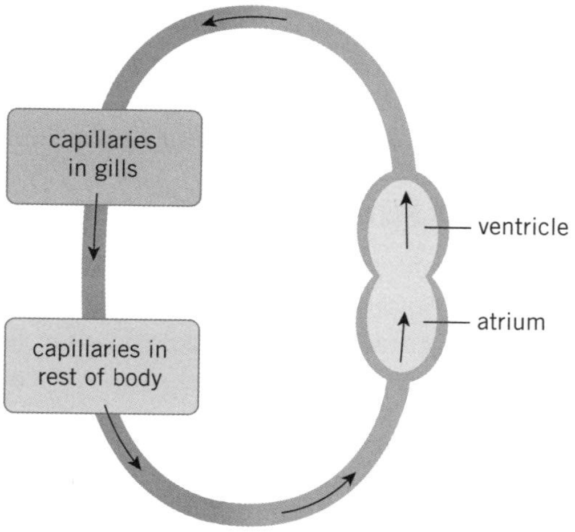	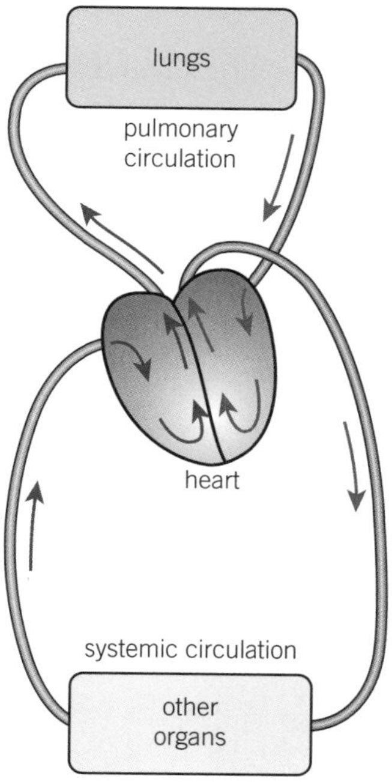
Description	Blood passes through the heart once per circulation	Two circuits; blood is pumped to the lungs and returns to the heart carrying oxygen. Blood is then pumped to body tissues before returning to the heart.
Examples	Fish, annelid worms	Birds, mammals

Summary questions

1 Flatworms lack a specialised circulatory system. As their name suggests, these animals have flattened body shapes. State and explain how nutrients and respiratory gases are transported in flatworms. *(2 marks)*

2 Explain the advantages of a closed double circulatory system over an open single circulatory system. *(3 marks)*

3 Fish possess closed single circulatory systems, but they are able to maintain high activity levels compared to other species with this type of system. Suggest why fish can be relatively active despite their single circulatory systems. *(3 marks)*

8.2 Blood vessels

Specification reference: 3.1.2(c)

You were introduced to the idea of closed circulatory systems in Topic 8.1, Transport systems in multicellular animals. Circulatory systems in animals include a variety of vessels, which are the subject of this topic.

The structures and functions of blood vessels

Blood travels through arteries after leaving the heart. Arteries branch into smaller arterioles, which diverge into capillaries. As blood travels back to the heart, capillaries converge to form venules, which lead into larger veins. The vessels' structural adaptations are suited to their positions in the circulatory system.

Arteries and veins consist of four layers: an inner **endothelium** layer, **elastic fibres**, **smooth muscle**, and **collagen** fibres. The proportions of each tissue present in vessels vary, depending on the requirements of the vessel. Capillaries, in contrast, have a single layer of endothelium cells.

Synoptic link

You learned about the structure of collagen and elastin in Topic 3.6, Structure of proteins.

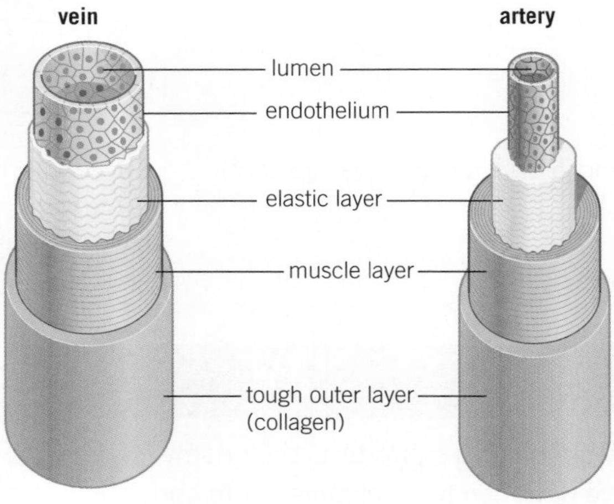

▲ **Figure 1** *Arteries and veins contain the same four tissues, but veins have wider lumens and less elastic tissue than arteries*

Revision tip: Blood vessel layers

The roles of the three outer layers in blood vessel walls can be summarised as follows:

Elastic fibres: provides flexibility

Smooth muscle: adjusts lumen size

Collagen: structural support

▼ **Table 1** *A comparison of blood vessel structure and function*

Vessel	Typical diameter	Relative proportion of elastin smooth muscle collagen	Key features	Explanation of key features
Arteries	5000 μm (0.5 cm)	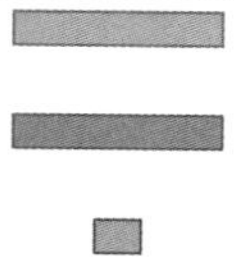	High proportion of elastic tissue	To stretch and recoil, which prevents rupture when the heart pumps
Arterioles	50 μm		Smooth muscle	The muscle contracts to narrow the lumen (vasoconstriction) and relaxes to widen the lumen (vasodilation). This controls where blood flows
Capillaries	10 μm	None	Thin walls (single layer of flattened endothelial cells, with gaps between them)	Permeable wall to enable diffusion of particles into tissue fluid (see Topic 8.3, Blood, tissue fluid, and lymph)
Venules	100 μm	Very little elastin or smooth muscle 	Thin walls (compared to veins)	Some permeability is retained, allowing the continued diffusion of some particles across the wall
Veins	10 000 μm (1 cm)		Wide lumen Valves, in most veins	Smooth blood flow at low pressure Backflow of blood is prevented

Summary questions

1 Explain how the structure of capillaries is suited to their function. *(3 marks)*

2 Explain the difference in elastic fibre content between arteries and veins. *(2 marks)*

3 Medium-sized veins can have diameters of 1 cm, whereas medium-sized arteries have diameters of approximately 0.5 cm. **a** Express these measurements in standard form. **b** Explain the difference in the typical diameters of arteries and veins. *(3 marks)*

8.3 Blood, tissue fluid, and lymph

Specification reference: 3.1.2(d)

Synoptic link

You learned about osmosis in Topic 5.5, Osmosis.

In Topic 8.2, Blood vessels, we examined how blood is transported in vessels. Blood delivers nutrients, oxygen, and chemical messengers to tissues, and removes waste products, such as CO_2. Yet this is only part of the story. Two other transport fluids exist in animals: tissue fluid and lymph. Here you will learn about the composition and roles of all three fluids.

Tissue fluid formation

Tissue fluid is formed when water diffuses out of blood capillaries, carrying dissolved solutes across the capillary wall. The key to tissue fluid formation is the balance between two pressures: osmotic and hydrostatic.

Along the length of a capillary, water potential is always higher in tissue fluid (and osmotic pressure is therefore higher in the blood). However, hydrostatic pressure changes along the length of a capillary.

- *At the arterial end of a capillary*: the hydrostatic pressure of the blood is high. This outweighs the higher water potential in the tissue fluid. Water diffuses out of the capillary.
- *At the venous end of the capillary*: the hydrostatic pressure of the blood is too low to outweigh the higher water potential in the tissue fluid. Water diffuses back into the capillary.

Cells are bathed in tissue fluid. This enables an exchange of materials: oxygen and nutrients enter cells, and waste products leave cells. Molecules such as hormones can also move between the tissue fluid and cells.

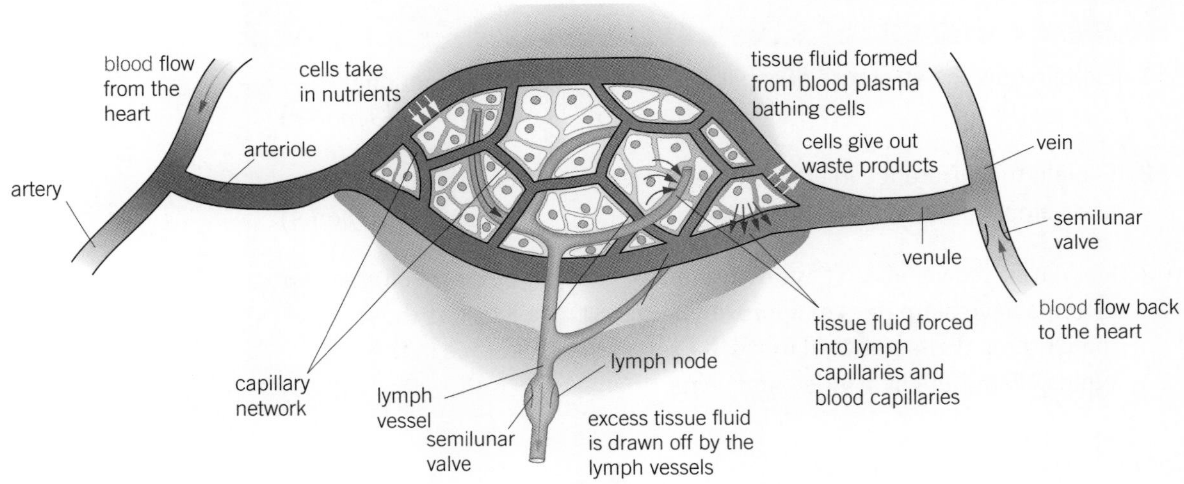

▲ **Figure 1** *The formation of tissue fluid*

Common misconception: Two pressures ... but can we consider one to be a 'pull'?

The direction that water moves though capillary walls is determined by the balance of two types of pressure: hydrostatic and osmotic.

Hydrostatic pressure = the pressure exerted by a fluid in confined spaces.

Osmotic pressure is more difficult to visualise. Water moves from an area of high water potential (with less dissolved solute) to an area of low water potential (with more solute). The solution with the lower water potential has the higher osmotic pressure; it will attract water via osmosis.

Considering osmotic pressure as a 'pull' is perhaps easier; osmotic pressure rises as solute concentration rises, which increases the probability of a solution attracting water.

Blood plasma in capillaries exerts a greater osmotic pressure than tissue fluid because it has more solute dissolved. In this case, proteins in blood (e.g. albumin) contribute significantly to the osmotic pressure. This represents a specific form of osmotic pressure called **oncotic pressure**.

Lymph

Some tissue fluid (approximately 10%) drains into the lymphatic system rather than re-entering the blood. Lymph fluid passes through lymph vessels, via lymph nodes, before returning to the bloodstream.

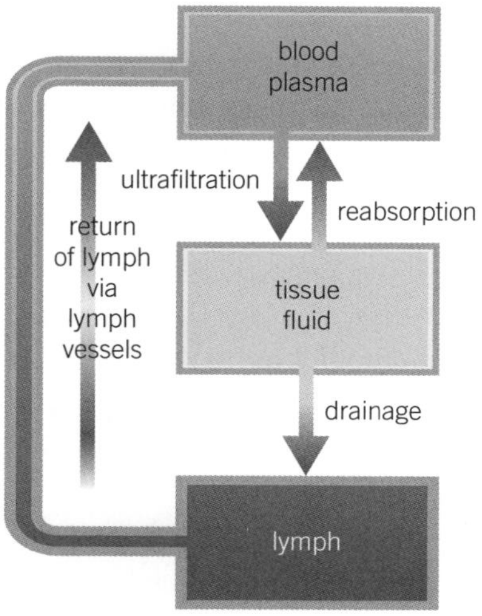

▲ **Figure 2** *The interactions between the three transport fluids*

Synoptic link

Topic 12.6, The specific immune system, provides details of the roles of leucocytes.

Functions and composition of the fluids

Fluid	Functions	Composition
Blood	Transport to and from tissues (the blood also plays a role in temperature regulation and as a pH buffer)	55% plasma (i.e. water plus solutes such as glucose, amino acids, ions, and large proteins, including hormones). 45% cells (leucocytes (white blood cells) and erythrocytes) and platelets.
Tissue fluid	Bathes cells, and exchanges materials with them	Few cells (which remain in the blood) other than some phagocytes (a type of leucocyte). Less solute than blood because many substances (e.g. oxygen, glucose, amino acids) diffuse into cells; no plasma proteins (e.g. albumin, fibrinogen).
Lymph	An important part of the immune system (e.g. phagocytes in lymph nodes ingest bacteria)	Similar to tissue fluid, but with less oxygen and nutrients, a greater proportion of fatty acids, and a large number of leucocytes.

Summary questions

1. State the differences between blood plasma and tissue fluid. (*3 marks*)
2. Explain the differences in the composition of tissue fluid and lymph. (*4 mark*)
3. Explain how tissue fluid is formed and reabsorbed into capillaries. (*5 marks*)

8.4 Transport of oxygen and carbon dioxide in the blood

Specification reference: 3.1.2(i) and (j)

Synoptic link

You read an overview of erythrocyte structure and function in Topic 6.4, The organisation and specialisation of cells. In Topic 3.7, Types of proteins, you learned about the structure of haemoglobin.

One of the functions of blood, as the transport medium in animals, is to deliver oxygen to respiring tissues and take away carbon dioxide. Here we examine how the transport and transfer of these respiratory gases is achieved in humans and other vertebrates.

Transport of gases in the blood

Oxygen transport

Oxygen binds to haemoglobin in red blood cells (erythrocytes). 98% of oxygen in the blood is transported by haemoglobin; 2% is carried in solution in the plasma. The reaction between haemoglobin and oxygen is reversible, which is crucial because O_2 is therefore released under the right conditions (see 'The Bohr effect', below).

$$\begin{array}{ccccc} \text{Hb} & + & 4O_2 & \rightleftharpoons & \text{Hb}(O_2)_4 \\ \text{Haemoglobin} & + & \text{oxygen} & \rightleftharpoons & \text{oxyhaemoglobin} \end{array}$$

Revision tip: Cooperation is the key

One haemoglobin molecule can bind with four O_2 molecules. One model of haemoglobin's chemistry suggests that the binding of the first O_2 molecule causes haemoglobin to change shape. This makes it easier for subsequent O_2 molecules to bind. This is known as positive cooperative binding.

Carbon dioxide transport

Small amounts of CO_2 are dissolved in blood plasma (5%) or combined with haemoglobin (10–20%). However, the majority (75–85%) is converted to HCO_3^- (hydrogen carbonate) ions.

$$CO_2 + H_2O \rightleftharpoons H_2CO_3 \rightleftharpoons HCO_3^- + H^+$$

Revision tip: Why is HCO_3^- ion production important?

The removal of carbon dioxide from plasma (and its conversion into HCO_3^- ions) maintains a steep CO_2 concentration gradient between respiring tissues and blood.

Chloride shift

The formation of HCO_3^- occurs in erythrocytes. HCO_3^- ions then move out of erythrocytes and into plasma. Chloride ions move into erythrocytes to replace them; this exchange is called the chloride shift.

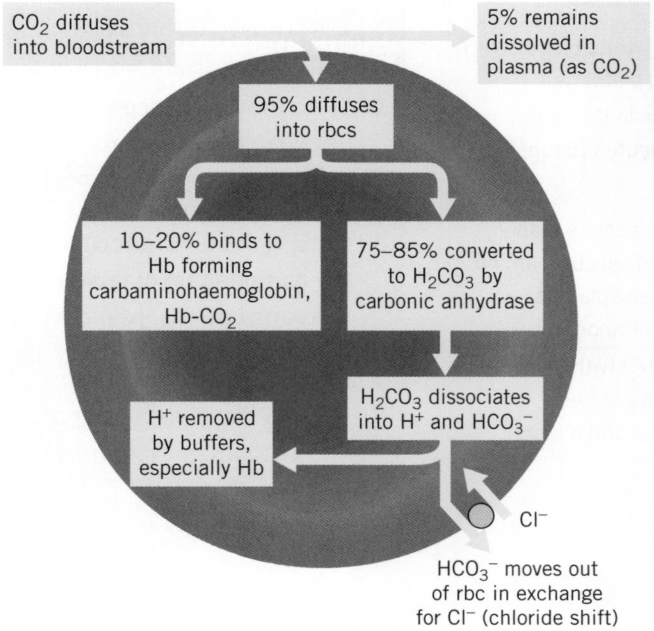

▲ **Figure 1** *Conversion of CO_2 to HCO_3^- ions is followed by the chloride shift (which transports HCO_3^- into the plasma)*

Oxygen dissociation curves

Oxygen dissociation curves illustrate how partial pressure affects the amount of oxygen binding to haemoglobin (i.e. % saturation). Small changes in oxygen partial pressure have a large influence on haemoglobin saturation.

The Bohr effect

Oxygen is released from haemoglobin more easily when carbon dioxide partial pressure (concentration) increases. This is known as the Bohr effect. CO_2 is produced in respiring tissues. This means O_2 will tend to be released where it is required, within respiring tissues. Conversely, CO_2 concentration in the lungs is low, which means O_2 binds more readily to haemoglobin in pulmonary capillaries.

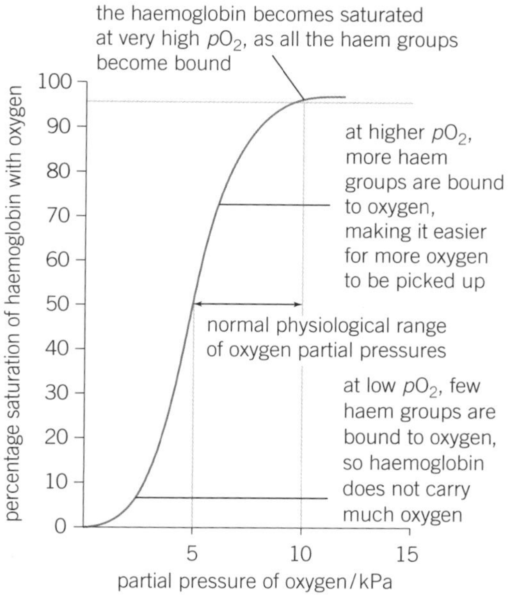

▲ **Figure 2** *The oxygen dissociation curve for adult haemoglobin*

Fetal haemoglobin

Fetal haemoglobin has a greater affinity for oxygen than adult haemoglobin (i.e. at a given partial pressure, fetal haemoglobin will be more saturated with oxygen than adult haemoglobin). This is important because it enables oxygen to be transferred from a mother's haemoglobin to a fetus's haemoglobin in the placenta.

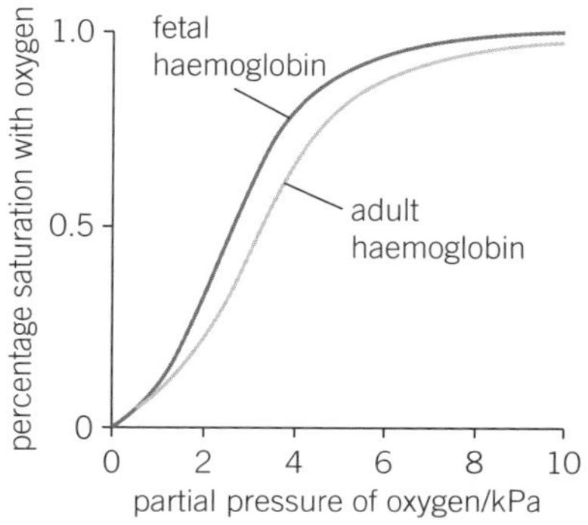

▲ **Figure 3** *Fetal haemoglobin binds to more oxygen than adult haemoglobin at a given partial pressure*

Go further: Other oxygen-carrying proteins are available ...

Haemoglobin is the molecule to which oxygen binds in the blood of vertebrates. Many **variants of haemoglobin** exist in humans and between species. Salt water crocodiles (*Crocodylus porosus*), for example, have been shown to produce haemoglobin with a different structure to haemoglobin in mammals and birds.

1 Mammals and birds are endotherms (animals that generate their own heat to maintain body temperature), whereas crocodiles are not. Suggest how the haemoglobin of crocodiles differs from that of endothermic species.

Myoglobin is an oxygen-binding protein found in vertebrate muscles. It has a higher affinity than haemoglobin for O_2, except at very high oxygen partial pressures. Myoglobin releases O_2 at very low oxygen partial pressure.

2 Suggest myoglobin's function in muscles, and sketch oxygen dissociation curves for myoglobin and haemoglobin on the same graph.

Oxygen is transported through some invertebrates by **haemocyanin**.

3 Haemocyanin is blue when oxygenated. Suggest a reason for the colour difference between haemoglobin and haemocyanin.

Revision tip: Partial pressures

The pressure exerted by a particular gas in a mixture is known as its partial pressure. In many circumstances, you can think of partial pressure as being equivalent to concentration.

Summary questions

1 **a** Describe the Bohr effect.
b Sketch a graph to show two oxygen dissociation curves: one at high CO_2 partial pressure and one at low CO_2 partial pressure. *(3 marks)*

2 Fetal blood exiting the placenta has an O_2 partial pressure of approximately 3 kPa. Use the graph in Figure 3 to estimate the difference in oxygen saturation between fetal and adult haemoglobin at 3 kPa. *(1 mark)*

3 Describe the shape of the oxygen dissociation curve for adult haemoglobin and explain the significance of the shape for human physiology. *(3 marks)*

8.5 The heart

Specification reference: 3.1.2(e), (f), (g), and (h)

Revision tip: Myogenic cardiac muscle

The muscle cells in the SAN pacemaker are myogenic, which means they can initiate muscle contractions themselves without nervous stimulation. Nervous stimulation from the brain can alter heart rate, however, by affecting the frequency at which the SAN initiates contractions.

Revision tip: The role of the bundle of His

Electrical activity passes down the bundle of His because this enables the ventricles to contract in the correct direction (i.e. the contraction begins at the apex, forcing blood up through the pulmonary artery and aorta).

The heart is the organ that pumps blood around an organism. Hearts range in complexity from a basic muscular tube in invertebrates to the four-chambered organ found in mammals. We focus on the mammalian heart in this topic.

Heart structure

The mammalian heart consists of four chambers. The right and left sides fill and empty together; they are divided by a septum.

▼ **Table 1** *The movement of blood in and out of the heart's chambers*

Heart chamber	Where does blood flow from when it enters the chamber?	Where does blood flow to when it leaves the chamber?
Right atrium	Deoxygenated blood from the vena cava	Right ventricle
Right ventricle	Right atrium (through atrioventricular valve)	Pulmonary arteries (through semilunar valve)
Left atrium	Oxygenated blood from pulmonary veins	Left ventricle
Left ventricle	Left atrium (through atrioventricular valve)	Aorta (through semilunar valve)

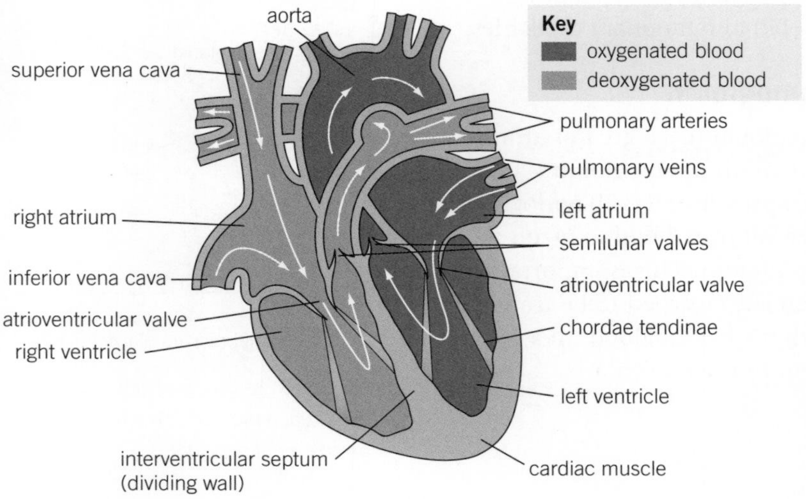

▲ **Figure 1** *The internal structure of the heart*

The cardiac cycle

The cardiac cycle comprises the events during one heartbeat. The cycle can be visualised in three stages: diastole (relaxation), atrial systole (the atria contract), and ventricular systole (the ventricles contract), which are illustrated below.

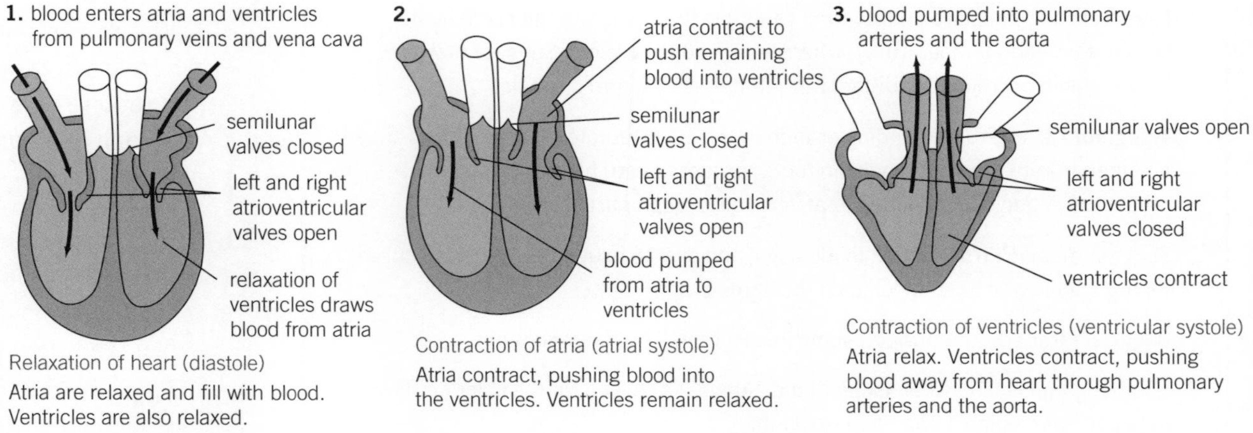

▲ **Figure 2** *The cardiac cycle*

What controls the rhythm of the cardiac cycle?

The cycle is coordinated through the following steps:

- The sino-atrial node (**SAN**), which is the heart's pacemaker, initiates a wave of electrical excitation.
- The atria are stimulated to contract. (*This is the atrial systole.*)
- The electrical impulse reaches the atrio-ventricular node (**AVN**).
- After a delay, the electrical activity passes down the **bundle of His** (made of Purkinje fibres).
- The ventricles contract from the apex (i.e. the bottom of the heart in most diagrams). (*This is the ventricular systole.*)

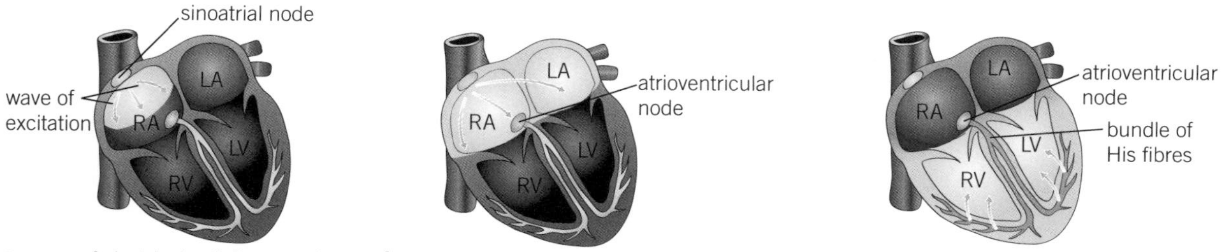

a wave of electrical activity spreads out from the sinoatrial node

b wave spreads across both atria causing them to contract and reaches the atrioventricular node

c atrioventricular node conveys a wave of electrical activity between the ventricles along the bundle of His and releases it at the apex, causing the ventricles to contract

▲ **Figure 3** *Control of the cardiac cycle*

Electrocardiograms (ECGs)

Electrocardiography can be used to monitor the electrical activity of a heart to check that a person's cardiac cycle is healthy. The results are shown on an ECG.

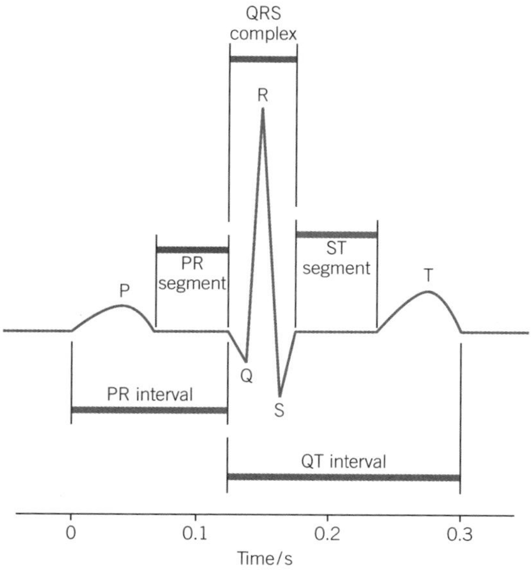

▲ **Figure 4** *An ECG trace for a single cardiac cycle. P corresponds to atrial systole; QRS represents ventricular systole; T is associated with diastole*

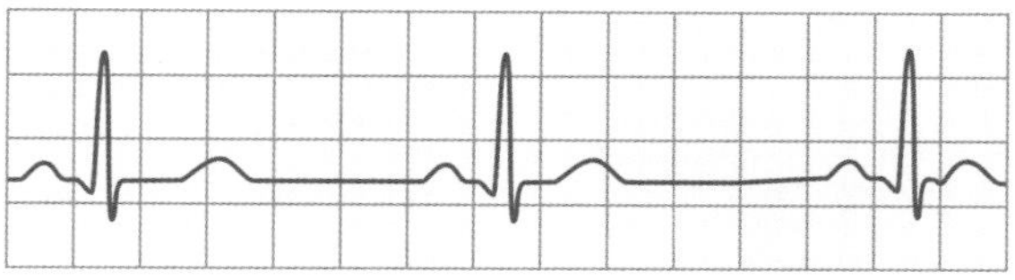

(b) Bradycardia – slow heart rate – beats evenly spaced, rate <60/min

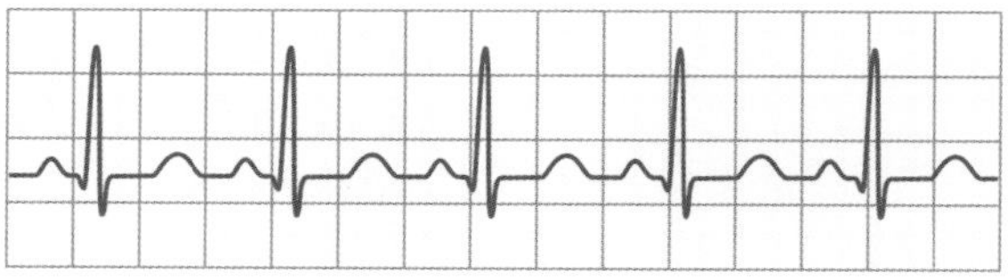

(c) Tachycardia – fast heart rate – beats evenly spaced, rate >100/min

▲ **Figure 5** *Two ECG traces showing abnormal heart rates*

Summary questions

1. Describe what occurs during **a** diastole **b** atrial systole **c** ventricular systole. *(3 marks)*
2. Explain why there is a delay between the P and QRS waves on an ECG trace. *(2 marks)*
3. Suggest what changes in atrial and ventricular pressure will occur through the cardiac cycle. *(3 marks)*

Chapter 8 Practice questions

1 Which of the following statements is true of double circulatory systems?

A They are open systems.

B They are found in some invertebrate species.

C They use haemolymph as a transport medium.

D They are found in bird species. *(1 mark)*

2 Which of the following statements is true of arteries?

A They possess an inner endothelium layer.

B Their lumens are always wider than the lumens of veins.

C They possess an outer elastic layer.

D Their elastic layers are narrower than those of veins. *(1 mark)*

3 Which of the following statements is true of lymph?

A It contains a higher concentration of oxygen than blood plasma.

B Lymph vessels contain valves.

C It contains no fatty acids.

D Lymph drains into tissue fluid. *(1 mark)*

4 Which of the following statements is true of carbon dioxide transport within the blood?

A Approximately 30% of CO_2 is dissolved in blood plasma.

B Carbon dioxide is converted to HCO_3^- ions in blood plasma.

C More oxygen binds to haemoglobin at high partial pressures of carbon dioxide.

D Approximately 15% of carbon dioxide in the blood binds to haemoglobin. *(1 mark)*

5 Which of the following terms describes a heart rate that is abnormally fast?

A Diastole

B Systole

C Tachycardia

D Bradycardia *(1 mark)*

6 The following passage describes the formation of tissue fluid. Complete the passage by choosing the most appropriate word to place in each gap.

Tissue fluid is formed when water in blood .... diffuses out of capillaries. .... pressure is higher at the arterial end of a capillary. Water .... remains higher in the tissue fluid along the length of a capillary. Water diffuses back into capillaries at the .... end. *(4 marks)*

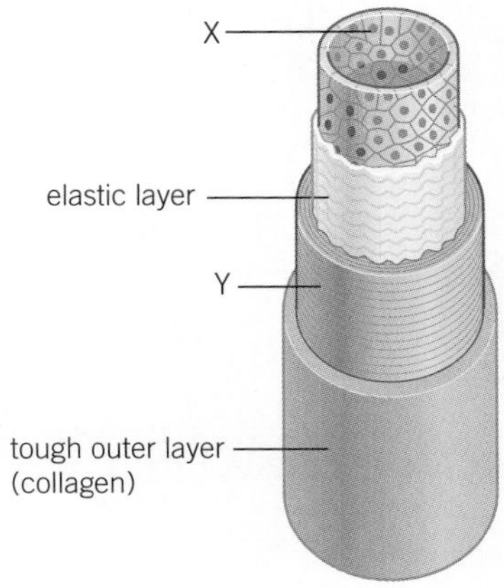

7 The figure to the left shows the generalised structure of a vein.

A Name the layers labelled X and Y. *(2 marks)*

B Describe the role of the elastic layer within blood vessels. *(2 marks)*

9.1 Transport systems in dicotyledonous plants

Specification reference: 3.1.3(a) and (b)(i–ii)

In Chapter 8 you learned about the transport systems that have evolved in animals. Plants also require internal systems to transport water and nutrients between roots, stems, and leaves.

Why do plants need transport systems?

As with multicellular animals, relying on diffusion alone to transport molecules and ions would fail to meet the demands of plants. They have high metabolic rates, can grow to remarkable sizes, and certain parts of plants (e.g. the trunks of trees) have small surface area to volume ratios.

Vascular systems

Vertebrate animals transport water and sugars within the same vessels. Plants, in contrast, employ separate vessels to carry water and minerals (**xylem** vessels) and sugars (**phloem**). These two tissues together are known as the vascular system. Vascular tissue has different arrangements in the roots, stems, and leaves of plants.

Revision tip: What is a dicotyledonous plant?

Dicotyledons are flowering plants with seeds that grow two primary leaves (cotyledons). As you may have guessed, monocotyledons (the other category of flowering plant) produce seeds that grow one primary leaf. Their vascular tissue has a different arrangement, which you are not required to learn.

Synoptic link

You can remind yourself about surface area to volume ratios in Topic 7.1, Specialised exchange surfaces. Topic 8.1, Transport systems in multicellular animals, outlines the circulatory systems that animals have evolved.

▼ **Table 1** *The arrangement of vascular tissue in different parts of a plant*

Roots (vascular system is central)	Stem (vascular system is peripheral)	Leaves
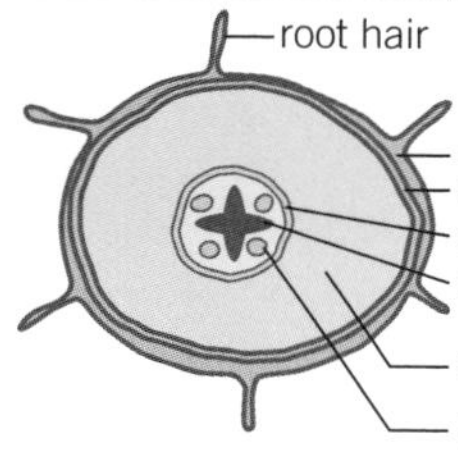root hair exodermis epidermis endodermis xylem cortex phloem	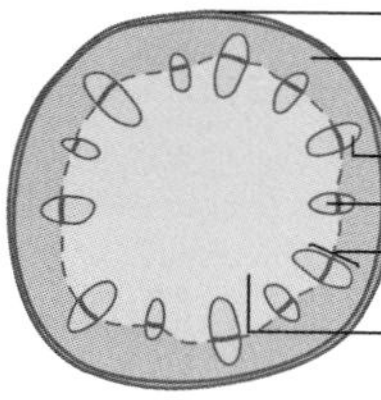epidermis cortex phloem xylem vascular bundle parenchyma (packing and supporting tissue)	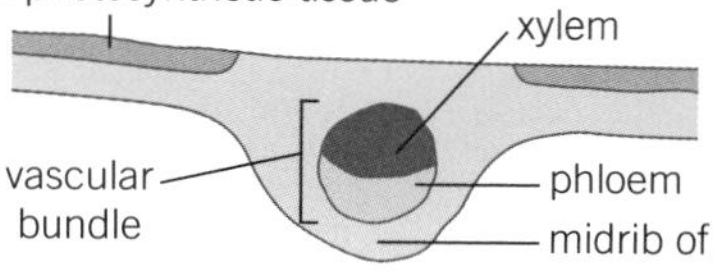palisade mesophyll– main photosynthetic tissue xylem vascular bundle phloem midrib of leaf

Xylem and phloem

▼ **Table 2** *The structure and function of xylem and phloem tissues*

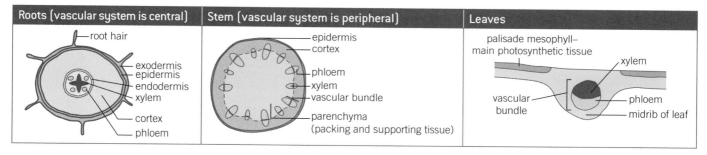

	Key features	Functions	Appearance
Xylem (see Topic 9.3, Transpiration, for more details of water movement up xylem vessels)	Dead cells have been fused to form hollow vessels, which are strengthened by lignin Pits in the xylem wall enable water to move out into adjacent xylem vessels or other cells Xylem fibres add strength, and xylem parenchyma cells store food	Transport of water and mineral ions up a plant The tissue also provides structural support for the plant.	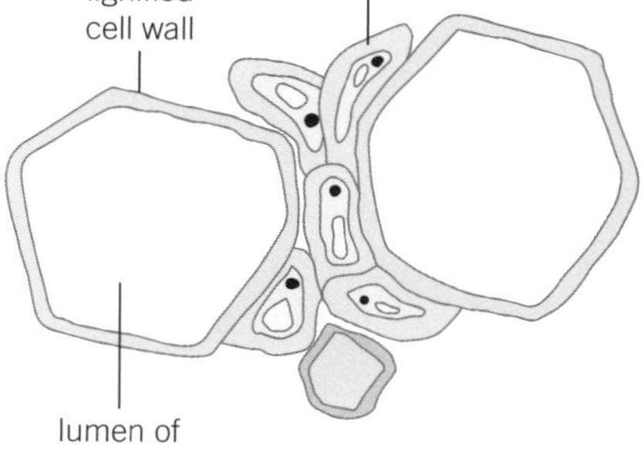xylem parenchyma lignified cell wall lumen of xylem vessel

Phloem (see Topic 9.4, Translocation)	Sieve tube elements are living cells joined end-to-end, forming a tube with internal pores (sieve plates) Companion cells carry out all the metabolic functions of the phloem tissue (as the tubes lack nuclei). Materials pass into sieve tubes via plasmodesmata (cell wall channels).	Transport of solutes (e.g. sugars and amino acids) up and down a plant	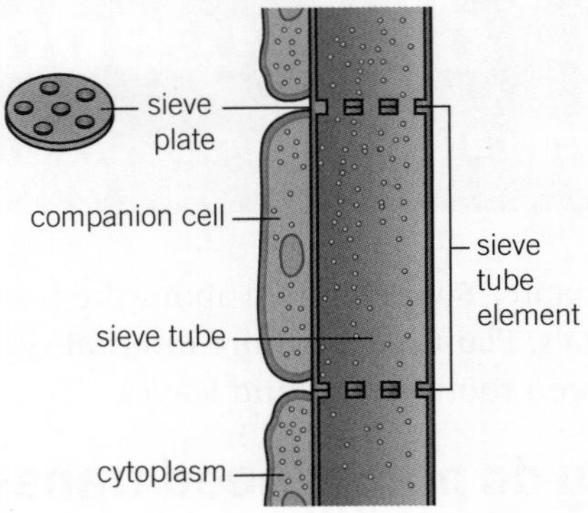

Revision tip: Dead or alive?

Xylem vessels and phloem sieve tubes both lack nuclei, but phloem sieve tubes are made from living cells, unlike xylem vessels.

Practical skill: examining plant tissue

Xylem vessels can be observed in either dead or living tissue. Plant stems or roots can be cut into thin sections, stained, and prepared on slides. The tissue is observed under a light microscope.

Xylem in some plant species (e.g. celery or gerberas) can be observed without the need to cut away tissue. The plants are first soaked in a solution of dye for at least 24 hours.

Summary questions

1 Describe the different arrangement of vascular tissue in roots and stems. *(2 marks)*

2 Outline how xylem vessels are adapted for water transport. *(3 marks)*

3 Explain how sieve tube elements survive, despite lacking nuclei and containing a small amount of cytoplasm. *(2 marks)*

9.2 Water transport in multicellular plants

Specification reference: 3.1.3(d)

Water plays numerous vital roles in plants, such as maintaining turgor, as a transport medium for mineral ions and sugars, and as a raw material in photosynthesis. Without a heart to pump fluid through its vessels, however, a plant relies on alternative mechanisms to transport water. You will learn about these mechanisms in the following two topics. We begin by looking at water movement from the soil to the xylem vessels.

Water uptake

Root hair cells are adapted to take up water from the surrounding soil because they are:

- long and narrow, which increases surface area to volume ratio
- able to penetrate between soil particles
- able to maintain a water potential gradient between the soil and the cell (due to solutes dissolved in the root).

Water moves into root hair cells by osmosis.

Movement towards the xylem

Water moves from cell to cell towards the xylem either through the cytoplasm (the **symplast** pathway) or through cell walls (the **apoplast** pathway). The flow of water is maintained by the *transpiration pull*, which you will learn about in Topic 9.3, Transpiration.

Entering the xylem

Water reaches the layer of cells (the **endodermis**) surrounding xylem vessels. An impermeable band (the **Casparian strip**) around endodermal cells forces water in the apoplast pathway back into the cytoplasm.

Endodermal cells move mineral ions into the xylem by active transport. As a consequence, the water potential in endodermal cells is higher than that of the xylem. Water diffuses into the xylem by osmosis. The initial flow into vascular tissue helps to force water up a stem; this is known as **root pressure**. The transpiration pull, which we discuss in the next topic, is a more significant factor in water movement than root pressure.

Question and model answer: Assessing evidence

Q. Outline the experimental evidence that supports the role of active transport in producing root pressure.

A. Active transport requires ATP, which is reliant on oxygen and carbohydrates (respiratory substrates). Root pressure drops when the concentrations of oxygen or carbohydrates are decreased. Cyanide prevents the production of ATP in respiration. The application of cyanide to root cells makes root pressure disappear. Root pressure increases as temperature is increased – this indicates that chemical reactions are likely to be occurring.

Summary questions

1 State two uses of water in plants. *(2 marks)*

2 Describe and explain the function of the Casparian strip. *(3 marks)*

3 Compare and contrast the movement of water through the apoplast and symplast pathways. *(5 marks)*

Revision tip: Water uptake – what's the solution?

The cytoplasm of root hair cells contains solutes such as sucrose and amino acids. Active transport of mineral ions from the soil further decreases the water potential inside the roots. This enables water to diffuse by osmosis down a water potential gradient from the soil.

Synoptic links

You read about the adaptations of root hair cells in Topic 6.4, The organisation and specialisation of cells.

The principles of water diffusion are outlined in Topic 5.5, Osmosis.

Revision tip: The Casparian strip

The Casparian strip is waterproof because it is composed largely of a waxy substance called suberin. The strip prevents toxic solutes from continuing to move into the plant, and it stops water from returning to the root cortex from xylem vessels.

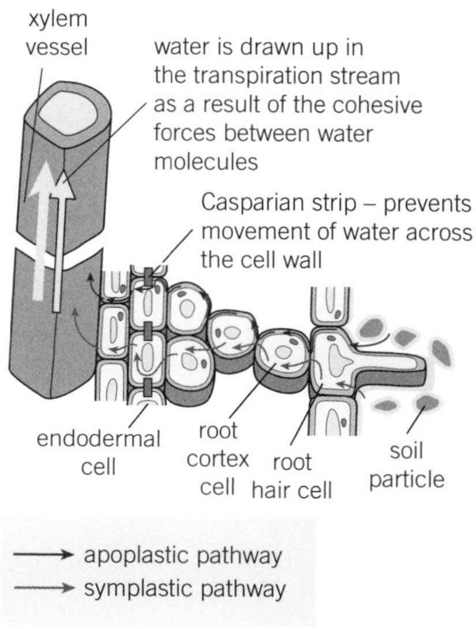

▲ **Figure 1** *Water movement from the soil to xylem tissue*

9.3 Transpiration

Specification reference: 3.1.3(c) and (d)

Synoptic link

The properties of water were discussed in Topic 3.2, Water.

Key term

Transpiration: The evaporation of water from a plant's leaves. (Water evaporates from cells inside the leaves and diffuses out of the stomata.)

Revision tip: A sticky subject

Try to avoid confusing *cohesion* (the attraction between water molecules) and *adhesion* (the attraction of water molecules to the inner surface of xylem vessels).

Synoptic link

You learned about the structure of guard cells in Topic 6.4, The organisation and specialisation of cells.

We examined how water enters plant roots in Topic 9.2, Water transport in multicellular plants. Here you will learn how water is moved through the rest of a plant.

Transpiration and water movement through xylem

- Water leaves a plant by **transpiration** (principally though stomata, which are discussed below).
- Water is pulled up through xylem vessels to replace the water lost through transpiration (i.e. the **transpiration pull**).
- Water molecules cohere to each other (i.e. they are attracted to each other through hydrogen bonding; they exhibit **cohesion**). This enables an unbroken chain of water molecules to be pulled up the xylem vessels (i.e. the **transpiration stream**).
- In addition, water molecules adhere to the sides of xylem vessels, which helps move the transpiration stream up the narrow vessels. This is called **capillary action**.

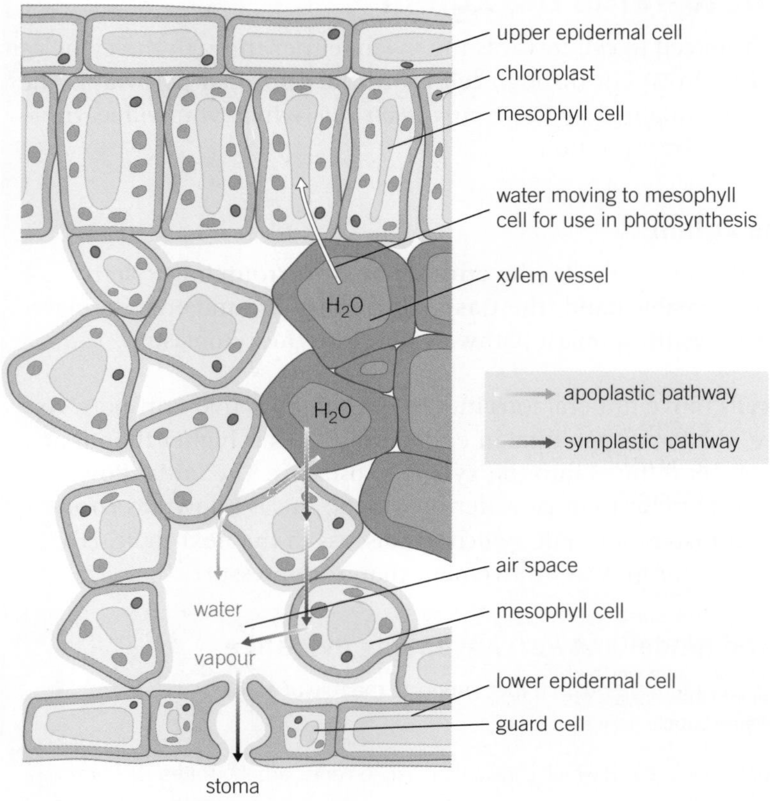

▲ **Figure 1** *Water movement within a leaf*

Evidence for the cohesion–tension theory

The model of water movement from soil to the leaves is known as the **cohesion–tension theory**. Much evidence has been gathered to support the theory, including:

- Trees become narrower when they transpire; this can be explained by the increased tension in xylem vessels during high rates of transpiration.
- Air is sucked up (rather than water leaking out) when a stem is cut.

- Water is no longer moved up a broken stem because the air that is pulled in breaks the transpiration stream (i.e. there is no longer a continuous chain of water molecules).

Stomata

In leaves, some water is used in photosynthesis, some exits through stomata (singular: stoma). The opening and closing of stomata is controlled by guard cells. When open, stomata allow the exchange of carbon dioxide and oxygen, but this also leads to transpiration and the loss of water.

Factors affecting transpiration

Both environmental factors and factors within a plant will affect transpiration rate, as shown in the following table.

Factor	How does it affect transpiration?	What increases transpiration rate?
Light intensity	Stomata open in the light	Higher light intensity
Temperature	Changes the kinetic energy of molecules	Higher temperature
Humidity (of the air)	Affects the water potential gradient between leaf and air	Lower humidity
Air movement	Affects how quickly moist air is removed	More air movement
Number of leaves	Affects the surface area available for loss of water vapour	More leaves
Number of stomata	Alters how much water is able to diffuse from the leaves	More (and larger) stomata
Thickness of cuticle	Waxy cuticles reduce water loss	Thin (or no) cuticles

Practical skill: measuring transpiration

Measuring water loss from plants is problematic. However, plants transpire approximately 99% of the water they absorb from the soil. Therefore measuring water uptake can give a good estimate of transpiration rates. Water uptake tends to be measured with a **potometer**.

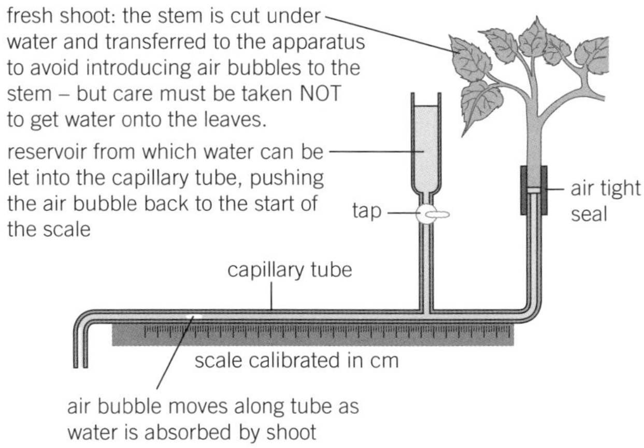

▲ **Figure 2** *A potometer*

The rate of water uptake is calculated by measuring the distance moved by the air bubble after a set time. Independent variables (e.g. temperature or light intensity) can be tested using a single plant species. Alternatively, water uptake in different species can be compared, which requires all environmental factors to be controlled (i.e. a constant temperature and light intensity).

Summary questions

1 State how each of the following conditions would affect transpiration rate. Explain your answers.
   a Decreased temperature
   b Increased light intensity
   c Absence of waxy cuticle. *(6 marks)*

2 Explain why the number of stomata in a leaf is likely to represent a compromise that depends on the environmental conditions in a plant's habitat. *(2 marks)*

3 The volume of water taken up by a potometer is calculated using this formula:
Volume = distance moved by bubble × πr^2 (where r = the radius of the capillary tube in the potometer)
If the mean distance moved by an air bubble is 12.0 mm in one minute, and the capillary tube has a diameter of 1.0 mm, calculate the rate of water uptake in $mm^3\ hr^{-1}$. *(4 marks)*

9.4 Translocation

Specification reference: 3.1.3(f)

Key term

Translocation: The movement of organic solutes through phloem sieve tubes.

Revision tip: Why transport sucrose?

Converting glucose to sucrose for translocation (and converting back to glucose at sinks) might seem inefficient. Glucose, however, is much more reactive than sucrose. This conversion prevents unwanted (and potentially harmful) reactions occurring during translocation.

Glucose produced in photosynthesis is converted to sucrose, which is transported around a plant. The transport of sucrose and other organic compounds occurs in phloem sieve tubes and is known as translocation.

Sources and sinks

Assimilates (the products of photosynthesis, such as sucrose) are translocated from sources to sinks. **Sources** include: green leaves and stems, and storage sites (such as tubers). **Sinks** include: developing seeds and fruits (which are laying down food stores), and growing roots.

The process of translocation

The movement of molecules into sieve tubes (phloem loading) can be an active or passive process (see Topic 9.2, Water transport in multicellular plants).

Passive loading: sucrose diffuses through cytoplasm and plasmodesmata (the symplast route) and enters sieve tubes.

Active loading: sucrose travels through cell walls and intercellular spaces (the apoplast route) and eventually reaches companion cells. The sucrose is loaded into companion cells through a combination of active transport and facilitated diffusion (as shown in Figure 1).

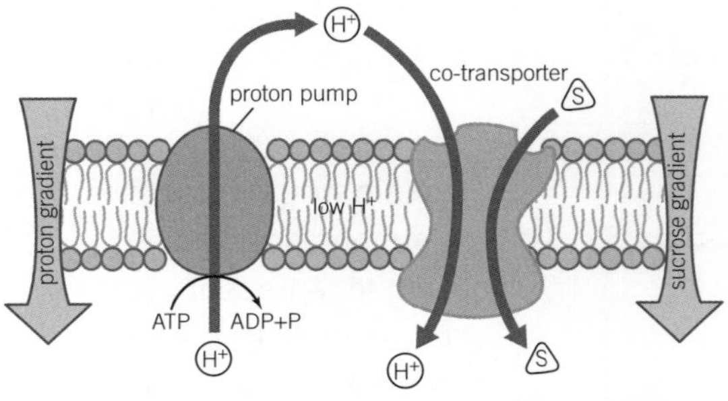

▲ **Figure 1** *The movement of sucrose into companion cells requires active transport*

In both cases, once sucrose is in sieve elements, water enters the phloem by osmosis. Turgor (water) pressure causes movement (mass flow) of water and dissolved solutes to regions of lower pressure. Water and solutes are unloaded from the sieve tubes at sinks (either by active or passive processes).

The evidence

Evidence for the mechanism of translocation includes: observation of the proteins required for active transport (thanks to advances in microscopy) the lack of translocation if mitochondria in companion cells are poisoned, and analysis of flow rates using aphid stylets.

Summary questions

1 State two examples of a translocation source and two examples of a sink. *(4 marks)*

2 Explain how sucrose is loaded into phloem cells using a combination of active transport and facilitated diffusion. *(4 marks)*

3 Suggest explanations for the distribution of carbohydrates in the parts of the plant shown in the following table. *(6 marks)*

Part of plant	Mean carbohydrate content ($\mu g\,g^{-1}$ fresh mass)			
	Sucrose	Glucose	Fructose	Starch
Leaf blade	1312	210	494	62
Vascular bundle in leaf stalk	5757	479	1303	<18
Tissue surrounding the vascular bundle in the leaf	417	624	1236	<18
Buds, roots, and tubers	2260	120	370	152

9.5 Plant adaptations to water availability

Specification reference: 3.1.3(e)

So far in this chapter you have examined how plants obtain, move, and use water. Some species, however, inhabit ecosystems in which water is scarce (**xerophytes**) or abundant (**hydrophytes**). Here you will learn about the adaptations evolved by both types of plant.

Xerophyte adaptations

Adaptation for conserving water	Benefit
Sunken stomata (in pits)	Traps moist air and reduces transpiration rate
Reduced number of stomata	Lowers the rate of transpiration
Reduction of leaf area (e.g. needle-like leaves in conifers)	Water loss via transpiration is reduced
Thick waxy cuticle	Reduces transpiration via the cuticle (which represents approximately 10% of water loss)
Curled leaves (e.g. in marram grass)	Traps moist air and reduces transpiration rate
Increased water storage (e.g. succulents such as cacti and samphire)	Provides a long-term reserve of water
Leaf loss during dry periods	Significantly reduces transpiration when no water is available
Long roots (e.g. some cactus' species)	Increases the chances of obtaining water from the ground

Hydrophyte adaptations

Adaptation for inhabiting aquatic ecosystems	Benefit
No waxy cuticle (or at least a very thin one)	The cuticle serves little purpose in hydrophytes because water loss is not an issue
Stomata tend to be open, and in plants with floating leaves stomata are on the upper surface	Gas exchange is maximised
A reduction in internal structural support	This would be another unnecessary feature because water can support aquatic plants
Air sacs and aerenchyma (specialised parenchyma tissue containing air spaces)	Increased buoyancy; leaves and flowers can float on the water surface
Small roots	Water uptake via roots is less significant because water can diffuse directly into other parts of submerged plants

Summary questions

1 List three ways in which xerophytes can be adapted to ecosystems with a lack of water. *(3 marks)*

2 Suggest how the stomata in hydrophytes might differ from those in land-based plants. Explain your answer. *(2 marks)*

3 Suggest why the results from a potometer experiment are not entirely representative of the transpiration rate of a plant in its natural habitat. *(3 marks)*

Key term

Adaptation: A trait that benefits an organism in its environment and increases its chances of survival and reproduction.

Synoptic link

You will learn more about the adaptations of marram grass in Topic 10.7, Adaptations.

Synoptic link

You read about stomata in Topic 6.4, The organisation and specialisation of cells, and Topic 9.3, Transpiration.

Chapter 9 Practice questions

1 The volume of water taken up by a plant in a potometer is calculated using this formula: volume = length of air bubble $\times \pi r^2$ (where r = radius of the capillary tube).

In an experiment, the mean distance moved by the air bubble in a potometer was 1.5 cm in one minute. The capillary tube had a diameter of 1.0 mm. Which of the following is the rate of water uptake in $mm^3\,hr^{-1}$?

A 11.8 **C** 706.5

B 70.7 **D** 2826.0 *(1 mark)*

2 Which of the following statements is/are true of the possible adaptations exhibited by hydrophytes?

1 Air spaces within stems

2 Rolled leaves to trap water vapour

3 Needle-shaped leaves with reduced surface areas

A 1, 2, and 3 are correct

B Only 1 and 2 are correct

C Only 2 and 3 are correct

D Only 1 is correct *(1 mark)*

3 Which of the following does water pass through in the apoplast pathway?

A Plasmodesmata **C** Vacuoles

B Cell walls **D** Cytoplasm *(1 mark)*

4 Which of the following statements is true of carbohydrate loading into companion cells?

A Sucrose is actively transported.

B Glucose is actively transported.

C Sucrose is co-transported.

D Glucose is co-transported. *(1 mark)*

5 Outline the evidence for the role of active transport in maintaining root pressure. *(3 marks)*

6

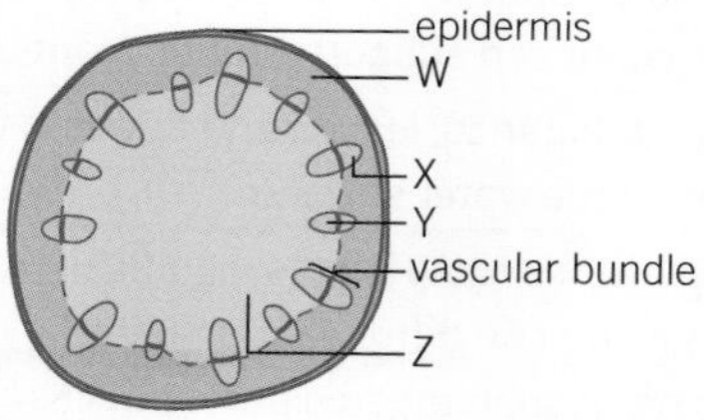

The figure shows the stem of a herbaceous plant.

Name the tissues found at W, X, Y, and Z. *(4 marks)*

7 Complete the following passage about the transpiration stream; choose the most appropriate word to place in each gap.

Water evaporates from .... cells in the leaves of plants. Water molecules diffuse out of stomata, down a concentration gradient. This lowers the water potential of cells within leaves. Water molecules move up .... Hydrogen bonds enable .... (i.e. water molecules being attracted to each other). Water moves up a plant's stem by .... action.

(4 marks)

10.1 Classification

Specification reference: 4.2.2(a) and (b)

Scientists estimate that more than 8 million species are alive on Earth today. Naming and grouping this vast array of organisms makes it easier for us to study them. The term we use for sorting organisms into groups is **classification**.

The classification system

Classification places organisms in taxonomic groups, or taxa (singular: taxon), which are organised in a hierarchy. The method scientists use to classify organisms is called the Linnaean system. Eight taxonomic levels are used, ranging from domains (the taxon containing the greatest number of organisms) to species (the smallest unit of classification). Table 1 illustrates the taxonomic groups of two species: common wheat and tigers.

▼ **Table 1** *The taxonomic groups of two eukaryotic species*

Taxonomic group	Organism	
	Common wheat	Tiger
Domain	Eukaryota	Eukaryota
Kingdom	Plantae	Animalia
Phylum	Magnoliophyta	Chordata
Class	Liliopsida	Mammalia
Order	Poales	Carnivora
Family	Poaceae	Felidae
Genus	*Triticum*	*Felix*
Species	*aestivum*	*tigris*

The number of species in each taxon increases as you move up the hierarchy. A genus (plural: genera), for example, contains at least one species, a family contains several genera, and an order comprises several families. The genetic similarity between organisms increases as you progress down the hierarchy from domain to species.

Advantages of a classification system

Classifying organisms into groups, using an agreed system, enables scientists to:

- analyse evolutionary relationships between organisms
- predict characteristics (as species grouped together are likely to share characteristics) and identify species
- share research findings (without confusion or ambiguity).

Key terms

Classification: The sorting of organisms into groups.

Species: A group of organisms that can interbreed to produce fertile offspring.

Revision tip: Mnemonics

Remembering the order of the taxonomic groups can be tricky. Inventing a mnemonic can help. One example is 'Don't Keep Pickled Cucumber Or Fried Gherkin Sauce'.

Revision tip: Naming species

The binomial system for naming species includes a few rules for you to remember:

1. The genus name begins with a capital letter.
2. The species name is always in lower case.
3. You should write both words in italics or, if handwritten, underline both words.

Summary questions

1. What is incorrect about each of the following binomial names?
   a *bufo fowleri* **b** Bufo americanus **c** *Turdus Merula* *(3 marks)*
2. Complete the Table 2 to show the classification of humans, *Homo sapiens*, and chimpanzees, *Pan troglodytes*. *(8 marks)*
3. The biological species concept defines a species as a group of organisms that can breed to produce fertile offspring. Suggest why this definition does not include all organisms and therefore may lack accuracy. *(2 marks)*

▼ **Table 2**

Taxon	Human	Chimpanzee
Domain		
Kingdom		
		Chordata
	Mammalia	Mammalia
	Primates	
	Hominidae	Hominidae
Genus		
Species		

10.2 The five kingdoms

Specification reference: 4.2.2(c)

Synoptic link

You learnt about the structure of prokaryotic and eukaryotic cells in Topic 2.4, Eukaryotic cell structure, and Topic 2.6, Prokaryotic and eukaryotic cells.

In Topic 10.1, Classification, you read about the principles of classification: how organisms are grouped together based on similarities. Here you will learn more about the taxonomic groups that have traditionally been considered the largest (the kingdoms). You will also consider a new level in the classification system (the domains) and the evidence that led to its development.

What are the five kingdoms?

The key features of the five kingdoms (Prokaryotae, Protoctista, Fungi, Plantae, and Animalia) are compared in Table 1.

▼ **Table 1** *A comparison of the features of the five kingdoms*

Kingdom	Unicellular or multicellular?	Domain(s)	Organelles present?	Cell wall	How do they gain nutrients?
Prokaryotae	unicellular	Bacteria and Archaea	No	Yes (peptidoglycan / murein)	autotrophic or heterotrophic
Protoctista	unicellular or multicellular	Eukarya	Yes	No	autotrophic or heterotrophic
Fungi	unicellular or multicellular	Eukarya	Yes	Yes (chitin)	saprotrophic (extracellular digestion)
Plantae	multicellular	Eukarya	Yes	Yes (cellulose)	autotrophic
Animalia	multicellular	Eukarya	Yes	No	heterotrophic

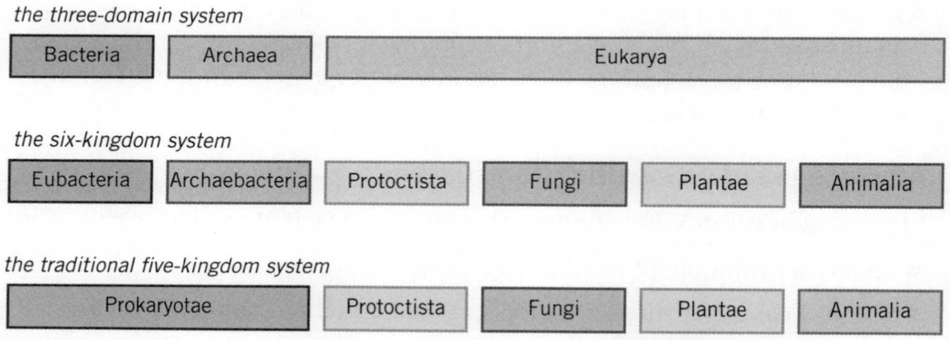

▲ **Figure 1** *The relationship between the three most common classification systems (three domains, six kingdoms, and five kingdoms)*

Why have domains been introduced?

A scientist called Carl Woese suggested dividing the Prokaryotae kingdom into two groups: Bacteria and Archaea. He placed the other four kingdoms (Protoctists, Fungi, Plants, and Animals) into a single group (the Eukarya, or Eukaryota). These three new groups were named domains.

Whereas early classification systems were based on observable anatomical features, modern systems rely on molecular comparisons between species. Woese's domain classification is based on several molecular observations, which include:

- Bacterial cell walls contain peptidoglycan but those of Archaea do not.
- The RNA polymerase of Archaea contains 8–10 subunits, but bacterial RNA polymerase contains only five subunits.
- Archaea have rRNA that is different to the rRNA of Bacteria and Eukarya.

Summary questions

1. State two differences between bacteria and fungi. *(2 marks)*
2. Classify the following species into the correct kingdom and domain.
   a. *Amoeba proteus* – a single-celled organism that lacks a cell wall.
   b. *Serpula lacrimans* – a species that uses extracellular digestion to decompose organic material.
   c. *Nanoarchaeum equitans* – a species discovered in 2002; their cell walls lack peptidoglycan, chitin, and cellulose. *(3 marks)*
3. Explain why classification systems have changed over time, using the introduction of domains as an example. *(4 marks)*

10.3 Phylogeny

Specification reference: 4.2.2(d)

In the previous topics you learned about how scientists classify organisms into groups that share particular features and evolutionary histories. The study of evolutionary relationships between species is called phylogeny. These relationships can be illustrated by constructing diagrams known as evolutionary (or phylogenetic) trees.

Key term

Phylogeny: The study of evolutionary relationships between organisms.

Producing phylogenetic trees

Whereas classification places species into discrete groups, phylogeny enables the construction of evolutionary trees, which show chronological relationships between species. These trees are not necessarily fixed. New evidence and different interpretations of evidence can result in the re-evaluation of evolutionary relationships. For example, a phylogenetic tree could be built from fossil evidence, but contradictory genetic evidence might indicate that an alternative phylogenetic tree is more appropriate.

Question and model answer: Interpreting phylogenetic trees

Q. Explain how the phylogenetic tree indicates which species is extinct.

A. Species E is extinct because the branch for this species ends before the present day.

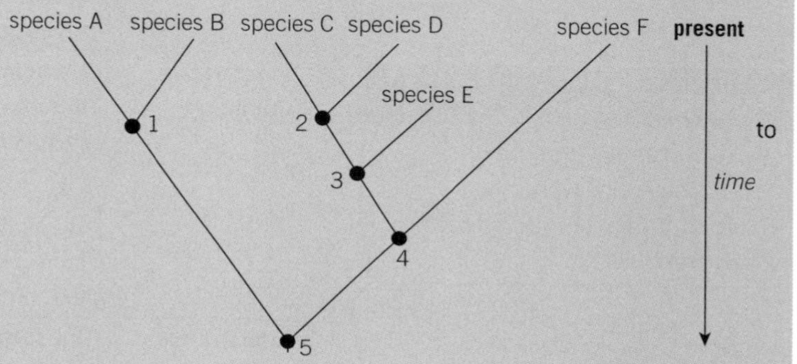

Q. Which species is most closely related to species A? Explain your answer.

A. Species B, which shares a common ancestor with species A at node 1. This indicates they evolved into separate species relatively recently.

Q. What does node 4 in this phylogenetic tree represent?

A. Node 4 represents a common ancestor of species C, D, E, and F.

Summary questions

1 Explain why the structure of a phylogenetic tree may change over time. *(2 marks)*

2 The phylogenetic tree below shows one interpretation of the evolutionary relationships between apes.

a Which was the first species in the tree to become extinct? *(1 mark)*

b Are gorillas more closely related to humans or gibbons? *(1 mark)*

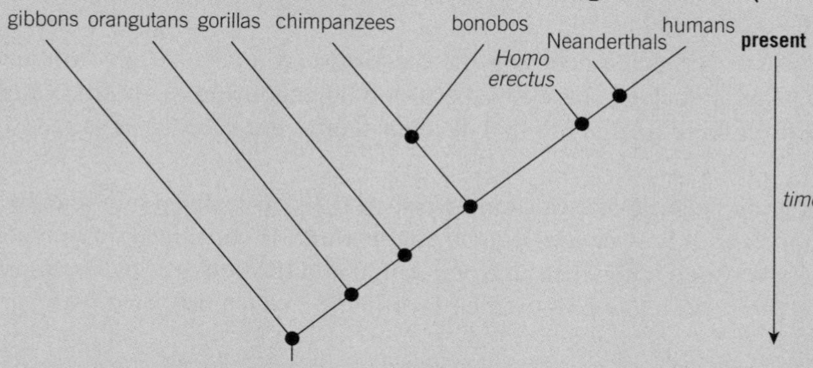

3 Suggest what advantages phylogenetic classification has over traditional hierarchical taxonomic classification. *(3 marks)*

Revision tip: Interpreting trees

You can interpret phylogenetic trees using the following rules:

- The earliest species is found at the base of the tree, and the most recent species are found at the tips of the branches.
- Branches for extinct species will end before the present day.
- The closer the relationship between two species, the more recently they will have branched from a common ancestor.
- Branch length is proportional to time.
- Nodes at branching points represent common ancestors.

10.4 Evidence for evolution

Specification reference: 4.2.2(e)

Phylogeny, which you studied in the previous topic, examines the evolutionary relationships between species. Here you will learn more about the theory of evolution, which explains how changes to species' characteristics occur over many generations.

Evidence for evolution

Evolution enables species to develop more advantageous phenotypes over time. These changes are caused by DNA mutations, which are inherited by subsequent generations. A wealth of evidence for evolution now exists, including the examples listed in the following table.

Type of evidence	What does the evidence show?
Palaeontology (fossils)	**Fossils** of simple organisms tend to be found in the oldest rocks, whereas more complex organisms are found in more recent rocks. Plant fossils appear in older rocks, before animals. This matches their ecological relationship (animals require plants to survive and will therefore have evolved later). Similarities between different fossil species (and extant species) reveal **gradual anatomical changes** over time.
Comparative anatomy	**Homologous structures** (anatomical features that have slight differences but the same underlying structure) provide evidence of divergent evolution (e.g. the vertebrate pentadactyl limb has evolved to perform a range of different functions in different species). The **embryos** of related species have similar appearances, despite considerable interspecific differences between adults. This indicates that different species have evolved from a common origin.
Comparative biochemistry	The rate of mutations in **DNA** can be calculated. This enables evolutionary relationships to be analysed. The closer the relationship between two species (i.e. the more recently their evolutionary paths diverged), the fewer the differences in their DNA base sequences (and therefore the fewer the differences in the primary structures of their proteins).

How was the theory of evolution developed?

Charles Darwin (in conjunction with Alfred Wallace) formed the initial theory of evolution by natural selection. Darwin was influenced by several ideas and observations, including:

- Charles Lyell's geological theories: fossils are evidence of animals living millions of years ago and natural processes can result from *gradual changes* and accumulations.
- His observations of finches on the Galapagos Islands: the slight differences (in beaks and claws) between species on neighbouring islands indicated that the birds were related but had developed different *adaptations* suited to their food sources.
- His studies of artificial selection by pigeon breeders enabled him to draw parallels with how *natural selection* might work. He concluded that *variation* exists between individuals in a population and the best adapted are more likely to survive and pass on their favourable characteristics to offspring.

Synoptic link

You will learn more about how natural selection works in Topic 10.8, Changing population characteristics.

Summary questions

1 What can the DNA base sequences of two species tell us about their evolutionary relationship? *(2 marks)*

2 Suggest why the fossil record fails to provide a complete picture of evolutionary history. *(3 marks)*

3 Explain the parallels between artificial selection and natural selection that Darwin observed. *(3 marks)*

10.5 Types of variation

Specification reference: 4.2.2(f)

In the previous topic we began to consider the principles of evolution: how species change over time. Evolution by natural selection can function only when there is variation between individuals in a population (intraspecific variation). The evolution of new species results in interspecific (between-species) variation. Here you will learn about the genetic and environmental causes of such variation.

The causes of variation

Variation between individuals can result from differences in their genetic material. Individuals of the same species can have different versions (alleles) of the same genes. Members of a species may have certain genes that other species lack. The environment introduces another layer of variation within populations. This is demonstrated by identical twins – they have the same DNA but will develop phenotypic variation during the course of their lifetimes based on differences in their environments.

Genetic variation

The following table outlines how genetic variation can be produced.

Cause of variation	How is variation introduced?	Synoptic link
DNA mutation	DNA base sequences are altered by mutations, producing new alleles.	2.1.3e Topic 3.9, DNA replication and the genetic code
Crossing over	Non-sister chromatids exchange genetic material during prophase I of meiosis, thereby creating new allele combinations.	2.1.6f Topic 6.3, Meiosis
Independent assortment	The random alignment of homologous chromosomes (metaphase I of meiosis) and sister chromatids (metaphase II) produces many allele combinations in gametes.	
Random fertilisation	During sexual reproduction, the identity of the two gametes that combine to form a zygote is largely due to chance.	

Only organisms that reproduce sexually experience crossing over, independent assortment, and random fertilisation. Mutations alone produce genetic variation in species that undergo asexual reproduction.

Environmental variation

Some traits are dictated solely by genetic variation (e.g. blood groups). Variation in characteristics is rarely determined by the environment alone, although physical damage (scarring) is one exception. Most variation results from a combination of genetics and the environment. The potential height to which a person can grow, for example, is governed by their genes, but their environment can influence whether or not they grow to this height (i.e. a poor diet may limit the person's height).

Revision tip: Intraspecific and interspecific variation

Variation between individuals in the same species is called in***tra***specific (you could remember this by thinking '***tra***pped within species'). This variation tends to involve minor differences (e.g. flower or fur colour, or height) and is a result of gene variants (i.e. members of a species have the same genes but can have different alleles of these genes).

Variation between members of different species is called in***ter***specific (*inter-* means 'between'; e.g. 'international' means 'between nations'). This variation tends to involve significant differences (although the closer the relationship between the species, the less significant the differences). Interspecific variation can be the result of species possessing different genes, as well as having different alleles.

Summary questions

1 State, with a reason, whether the following statements describe interspecific or intraspecific variation. **a** Arctic rose plants (*Rosa acicularis*) usually have pink flowers, but some have white flowers. **b** The shape of arctic rose (*Rosa acicularis*) leaves is pinnate. The desert rose (*Rosa stellata*) has trifoliate leaves. **c** Oystercatchers (*Haematopus ostralegus*) and zebra finches (*Taeniopygia guttata*) both have bills that vary in colour from orange to red. *(3 marks)*

2 Describe how genetic variation can be introduced into a species. *(4 marks)*

3 This graph shows some different conditions shared by twins and illustrates the relative influence of genes and the environment on each trait. Explain the evidence for the comparative influence of genetics and the environment on height and strokes. *(3 marks)*

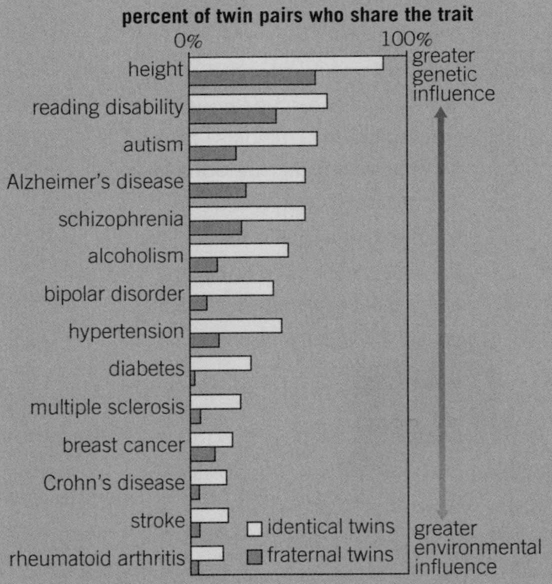

10.6 Representing variation graphically

Specification reference: 4.2.2(f)

We began thinking about biological variation in Topic 10.5, Types of variation. Variation within and between populations can be displayed and analysed by scientists in order to reveal patterns, make comparisons, and draw conclusions. Here you will learn about three statistical analysis techniques: standard deviation (for measuring the extent of variation in a population), Student's *t* test (for comparing data between two populations), and correlation coefficients (for analysing whether a correlation exists between two variables).

Discontinuous vs continuous variation

The type of data scientists collect determines how it should be represented graphically. Data can show either continuous or discontinuous variation. The features of both types of variation are outlined in the following table.

	Discontinuous variation	Continuous variation
Nature of the data	Discrete values with no intermediate values	A range of values (i.e. a continuum)
Genetic influence	One gene or a small number of genes	Several genes (polygenic)
Environmental influence	No	Yes
How is it represented?	Bar chart / pie chart	Histogram
Examples	Gender, blood groups, bacterial shape, whether flowering plants are monocots or dicots	Height, weight, number of pollen grains produced, rate of binary fission

Representing data for continuous variation

A trait that shows continuous variation often produces a normal distribution curve when population data are displayed in a graph. A **normal distribution curve** is a symmetrical, bell-shaped curve where most values lie close to the mean.

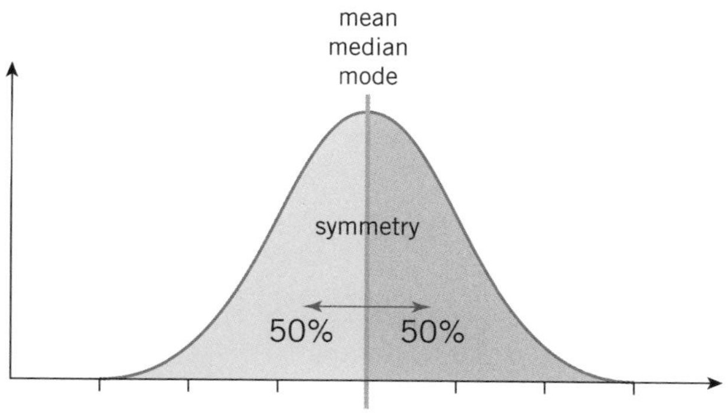

▲ **Figure 1** *A normal distribution curve*

Standard deviation

Standard deviation is calculated to assess the variation within a data set. A high standard deviation value indicates that the data shows a lot of variation. The calculation of standard deviation is performed on data with a normal distribution.

Worked example: Standard deviation

Standard deviation (σ) is calculated as:

$$\sigma = \sqrt{\frac{\sum(x-\bar{x})^2}{n-1}}$$

$\sum$ = the sum (total) of

x = value measured

$\bar{x}$ = mean value

n = total number of values in the sample

Here is a worked example of the calculation, using the horn length of topi antelopes as the data set:

Sample number	1	2	3	4	5	6	7	8	9	10
Horn length (cm)	38.4	39.0	39.2	40.1	39.4	38.9	40.2	39.2	40.8	40.8

1 First, calculate the mean ($\bar{x}$) by dividing the sum of the measurements by the number of individuals sampled. In this case, the calculation is 396 / 10 = 39.6 cm

2 The mean value is then subtracted from each of the 10 measured values for horn length ($x - \bar{x}$). For example, 38.4 – 39.6 = –1.2. This gives us a mixture of positive and negative values: –1.2, –0.6, –0.4, 0.5, –0.2, –0.7, 0.6, –0.4, 1.2, 1.2

3 These differences are then squared $(x - \bar{x})^2$ (e.g. $-1.2^2 = 1.44$) and the squared differences are added together ($\sum(x - \bar{x})^2$). In this case, the calculation is 1.44 + 0.36 + 0.16 + 0.25 + 0.04 + 0.49 + 0.36 + 0.16 + 1.44 + 1.44 = 6.14 cm.

4 The sum of the squared differences ($\sum(x - \bar{x})^2$) is divided by the sample size minus one. In this example, the calculation is 6.14/9 = 0.682.

5 We then find the square root of this value to calculate the standard deviation. $\sqrt{0.682} = 0.826$

What does this σ value of 0.826 mean? It tells us that 68% of all measured values should be within 1 standard deviation of the mean. If we measured every topi in the population, we would expect 68% of individuals to have horn lengths in the range 38.774 to 40.426 cm (i.e. 39.600 cm ± 0.826). Furthermore, 95% of individuals should be within 2 standard deviations of the mean (i.e. 95% of topis should have a horn length between 37.948 and 41.252 cm).

Student's *t* test

Two sets of data can be compared using a *t* test. The calculation assesses whether there is a difference between the two data sets.

The formula for the *t* test is:

$$t = \frac{(\bar{x}_1 - \bar{x}_2)}{\sqrt{\left(\frac{\sigma_1^2}{n_1}\right) + \left(\frac{\sigma_2^2}{n_2}\right)}}$$

$\bar{x}_1, \bar{x}_2$ = mean of populations 1 and 2

σ_1, σ_2 = standard deviation of populations 1 and 2

n_1, n_2 = total number of values in samples 1 and 2

The null hypothesis and *t* value

The *t* test tells you whether a significant difference exists between two sets of data. The **null hypothesis** being tested is that no difference exists. Once you have calculated your *t* value, you need to:

1 Work out the **degrees of freedom** in the test (this is calculated as the total number of samples from both data sets, minus 2 ($n_1 + n_2 - 2$)).

2 Use a table of probabilities (which you will be given in your exam) to find a **probability value** for your *t* value, corresponding to the correct degrees of freedom.

3 The probability value tells you the likelihood of the difference between the two data sets being due to chance.

4 You can reject your null hypothesis if there is less than a 5% probability that the difference is due to chance. A probability (p) value of 0.05 or less tells you that there is a **significant difference** between the two groups you have been investigating.

Spearman's rank correlation coefficient

A correlation coefficient (r_s) is calculated to assess whether a relationship exists between two sets of rank-ordered data. The formula for calculating r_s is:

$$r_s = 1 \frac{6\Sigma d^2}{n(n^2 - 1)}$$

where:

r_s = correlation coefficient

Σ = the sum (total) of

d = difference in ranks

n = number of pairs of data

The value of r_s will be between −1 and +1; a perfect negative correlation produces a value of −1, whereas a perfect positive correlation gives a value of +1. You will need to look up a probability (p) value for your r_s value (using the same procedure as you would with a *t* value) in order to conclude whether a correlation is significant. The critical values table for a Spearman's rank test is different from that of a *t*-test so you must check you are using the correct one.

Revision tip: Understand the formulae rather than memorising them

You are not expected to learn any of these formulae. However, you will be required to use them, if they are provided, and to understand what the results indicate.

Summary questions

1 State whether the following examples show discontinuous or continuous variation: **a** the volume of a protozoan cell **b** the presence or absence of haemophilia in a person **c** resting heart rate. *(3 marks)*

2 Horn lengths were measured in two populations of topi.

	Number sampled	Mean horn length (cm)	Standard deviation (cm)
Population 1	10	41.200	0.930
Population 2	10	39.600	0.826

Use the *t* test to assess whether the two populations show a significant difference in horn length.
For 18 degrees of freedom, a *t* value of 2.10 corresponds to a *p* value of 0.05. *(5 marks)*

3 The mating success of 10 male topis was recorded along with their horn lengths.

Sample number	1	2	3	4	5	6	7	8	9	10
Horn length (cm)	38.400	39.000	39.200	40.100	39.400	38.900	40.200	39.200	40.800	40.800
Mean number of females mated per day	0.800	1.333	1.250	2.200	1.250	1.500	2.500	1.500	2.000	1.800

Calculate a Spearman's rank correlation coefficient to assess whether mating success is correlated with horn length in this population.
For 18 degrees of freedom, an *r* value of 0.40 corresponds to a *p* value of 0.05. *(6 marks)*

10.7 Adaptations

Specification reference: 4.2.2(g)

In the last two topics, you have explored variation between organisms. Any variation that benefits an organism in its environment is called an adaptation. Here you will learn about the types of adaptations exhibited by species. You will also look at situations in which unrelated species develop similar adaptations, which is known as convergent evolution.

Types of adaptation

Adaptations fall into three categories: anatomical, behavioural, and physiological.

Type of adaptation	Description	Roles of the adaptations: examples
Anatomical	Physical structures	***Communication*** (e.g. species such as the flamboyant cuttlefish (*Metasepia pfefferi*) display bright colours to warn other species of their toxicity). ***Locomotion*** (e.g. fish species have evolved gas-filled swim bladders for buoyancy, fins, and streamlined shapes). ***Feeding*** (e.g. the structure of animals' teeth have evolved to suit their diets; herbivores have cusped molars for grinding whereas carnivores require sharp canines for tearing meat). ***Water regulation*** (e.g. some plant species, such as marram grass, need adaptations (thick waxy cuticles, curled leaves) to conserve water).
Behavioural	Simple, innate behaviours (such as reflexes) through to more complex (learned) behaviours	***Communication*** (e.g. some animal species use elaborate courtship displays to attract mates). ***Locomotion*** (e.g. taxes, which are simple behaviours involving organisms moving in one direction, either towards or away from a stimulus; *E. coli* bacteria, for example, exhibit chemotaxis when they are attracted to certain chemicals). ***Responding to seasonal changes*** (e.g. migration of certain animal species and hibernation of mammalian species such as ground squirrels).
Physiological	Biochemical and cellular traits (e.g. the type of enzymes and hormones an organism produces)	***Feeding*** (e.g. species have evolved different enzymes to digest the components of their diets). ***Antibiotic resistance*** (bacteria can evolve resistance to antibiotic drugs though a variety of molecular adaptations). ***Adaptations to temperature*** (e.g. icefish (*Chaenocephalus aceratus*) produce cold-resistant enzymes that are suited to their Antarctic climate).

Convergent evolution

Unrelated species can live in similar habitats and face similar selection pressures. This results in such species evolving, independent of each other, similar structures (known as analogous structures). This is called convergent evolution (because the phenotypes of unrelated species converge). Examples of convergent evolution include:

- Leaves, which have evolved in plants on several separate occasions.
- The ability to produce silk threads, which has evolved in weaver ants, spiders, and silk moths.
- Whales (order: Cetacea) and dugongs (order: Sirenia), which have evolved similar tail flukes.
- Echolocation (for hunting), which has evolved in bats, whales, and oilbirds.

Common misconception: Adaptation takes time

A species adapts over many generations. Adaptation is the process by which species evolve traits to suit their habitat. This tends to be a lengthy process, often taking millions of years. To speak of individual organisms *becoming* adapted is incorrect. Individuals *exhibit* adaptations (i.e. traits that have evolved over many generations).

Summary questions

1 State whether the following adaptations should be classified as anatomical, behavioural, or physiological: **a** sweating **b** phototaxis **c** opposable digits **d** lactose tolerance. *(4 marks)*

2 Aye-aye lemurs (Order: Primates) and striped possums (Order: Diprotodontia) both have elongated fingers that they use to locate invertebrates in trees. Explain why this is an example of convergent evolution. *(3 marks)*

3 Outline the anatomical, behavioural, and physiological adaptations of bacterial cells. *(6 marks)*

Synoptic link

You learned about plant adaptations to prevent water loss in Topic 9.5, Plant adaptations to water availability.

Key term

Adaptation: A trait that benefits an organism in its environment and increases its chances of survival and reproduction.

10.8 Changing population characteristics

Specification reference: 4.2.2(g) and (h)

You have examined some of the evidence for evolution by natural selection in Topic 10.4, Evidence for evolution, and in the previous topic you learned about the types of adaptation that species can evolve. Here you will delve further into the mechanism of natural selection. You will also consider some recent examples of evolution and their implications for humans.

Natural selection

The characteristics of a population evolve over time through a mechanism called natural selection. The following steps are required:

1. **Variation** (due to genetic mutations) must exist within the population.
2. The presence of a **selection pressure** (a factor that affects an organism's chance of survival, such as the threat of predation, diseases, climate change, and changes in food availability).
3. Only individuals possessing traits enabling them to overcome the selection pressure will survive and reproduce (because they are better **adapted** than other organisms).
4. The gene variants (**alleles**) that enabled survival in the presence of the selection pressure are passed on to the next generation.
5. Over many generations the allele frequencies in the population will change and the balance of phenotypic characteristics will change accordingly to suit the population's habitat.
6. When a series of mutations arises in a population, natural selection can result in **speciation** (the formation of a new species).

Go further: A case of speedy evolution

Deer mice (*Peromyscus maniculatus*) are widespread across the USA. Most deer mice have a dark coat. However, in an area of Nebraska known as Sand Hills, a deer mouse population has evolved lighter fur to match the sandy soils in this habitat. The remarkable part of this story is that Sand Hills only formed 8000–15 000 years ago. The evolution of the deer mouse coat happened within the space of a few thousand years.

Researchers estimate that the allele for a sandy-coloured coat arose 4000 years ago. This produced variation in fur colour within the population. The deer mice would have faced a selection pressure in the form of predation (principally from birds). Mice with sandy fur would have been better camouflaged and less likely to be eaten. These individuals would have been better adapted than those with dark fur. Sandy mice would have been more likely to survive, reproduce, and pass on their alleles to offspring.

Over many generations, the frequency of the allele for sandy fur would have increased until the population comprised mainly sandy-coloured mice. The researchers calculated that the sandy coat allele gave these mice a 0.5% survival advantage.

1. Explain why this is an example of natural selection but not speciation.
2. Suggest why the allele frequencies for fur colour changed over time despite the sandy coat allele giving the mice only a 0.5% survival advantage.

Modern evolution: implications for humans

Some species evolve at a faster rate than others. The recent evolution of some species has implications for human populations. Some examples are outlined in the following table.

Species	Adaptation that has evolved	Implication for humans
Staphylococcus aureus	Antibiotic resistance (e.g. to methicillin)	Methicillin-resistant *Staphylococcus aureus* (MRSA) infections are very difficult to treat.
Flavobacterium sp	Production of nylonase enzyme	These bacteria can be used to remove factory waste.
Drosophila sp	Resistance to the insecticide malathion	*Drosophila* fruit flies can infest orange groves.

Summary questions

1 State three examples of a selection pressure that bacterial species may encounter. *(3 marks)*

2 Describe one example of an adaptation evolved in another species that has benefited humans. *(2 marks)*

3 Humans have developed chemicals called insecticides to kill undesired insect species. Explain how a species of insect could become resistant to an insecticide. *(4 marks)*

Chapter 10 Practice questions

1 Which of the following is an example of a physiological adaptation in humans?

A Tool use
B Bipedalism
C Lactose tolerance
D Opposable digits *(1 mark)*

2 Which of the following is a feature of prokaryotic organisms?

A Membrane-bound organelles
B Chitin cell wall
C 80S ribosomes
D Circular DNA *(1 mark)*

3 For which of the following examples would the Student's *t* test be an appropriate statistical test to use?

A Measuring the variation in trunk length in an African elephant (*Loxodonta africana*) population.

B Comparing the mean fat reserves of two populations of grizzly bears (*Ursus arctos*).

C Analysing the relationship between temperature and leaf length in common ivy (*Hedera helix*). *(1 mark)*

4 The following passage concerns the process of natural selection. Complete the passage by choosing the most appropriate word to place in each gap.

.... can produce new gene variants, known as .... Some gene variants code for traits that provide a survival advantage when a population experiences a .... (e.g. a new disease or a change in climate). Individuals that are best .... to their environment will have a higher probability of surviving and passing on their genes to the next generation. *(4 marks)*

5 The graph below shows the range of heights in a sample of a human population.

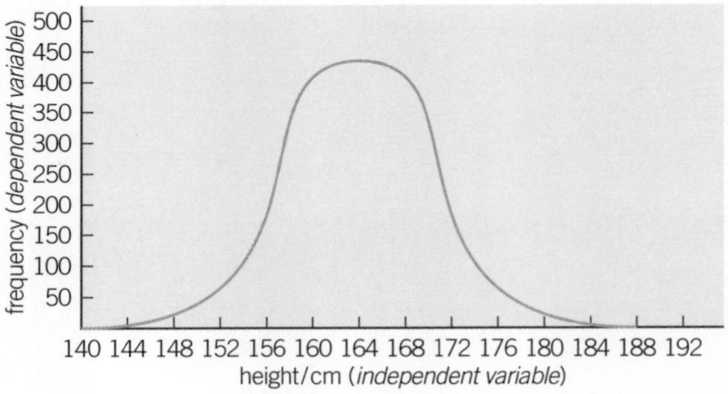

a State the name given to this type of distribution. *(1 mark)*

b **i** What type of variation is shown in the graph? *(1 mark)*

ii Explain why height exhibits this type of variation. *(2 marks)*

c Estimate the mean, mode, and median height values from the graph. *(2 marks)*

6 Describe the evidence that can be used to classify species.

You should illustrate each type of evidence you include in your answer with examples. *(6 marks)*

11.1 Biodiversity

Specification reference: 4.2.1(a)

Biodiversity, in a broad sense, is the variety of life in an area. We can, however, consider biodiversity on several scales: the genetic variation within one species, the variety of species in a habitat, or the range of habitats in an area.

Measures of biodiversity

	Genetic diversity	Species diversity	Habitat/ecosystem diversity
What is measured?	An estimate of gene variants (alleles) in a species.	Two components are measured: **Species richness** (the number of species in an area). **Species evenness** (the number of individuals of each species).	The number of different habitats (or ecosystems) in an area. This is the hardest measure of biodiversity to calculate because the boundaries of ecosystems are often difficult to determine with accuracy.
Topic reference	11.5, Calculating genetic biodiversity	11.4, Calculating biodiversity	11.6, Factors affecting biodiversity

Key term

Biodiversity: The variety of life in an area, which can be measured in terms of genetic, species, or habitat diversity.

Common misconception: Ecological terms

In this chapter you will be wading through several related ecological terms, which are easy to confuse. For example, a **population** is a group of organisms of the *same* species. A **community** is a collection of populations in a **habitat**, which is the natural environment of organisms. An **ecosystem** is formed from a community of species (the **biotic** component) and their non-living surroundings (the **abiotic** components such as air, water, and soil) interacting together.

Revision tip: Richness and evenness

An ecosystem is considered biodiverse only when species richness and evenness are both high. For example, many species may be identified in a habitat, indicating species richness. Yet if one or two of the species dominate then species evenness is low.

Summary questions

1 Describe the difference between a population and a community. *(2 marks)*

2 Suggest and explain which measure of biodiversity is most significant when assessing the health of one ecosystem. *(2 marks)*

3 Explain how a habitat can contain many species but its biodiversity is considered low. *(3 marks)*

11.2 Types of sampling

Specification reference: 4.2.1(b)

Ecologists can monitor biodiversity by sampling parts of an ecosystem. Counting every single organism is impossible, but calculating the biodiversity in small sections of an ecosystem enables the overall biodiversity to be estimated.

The principles of sampling

The choice of sampling method is an important decision. You should bear in mind a few key principles when designing your sampling method.

- **Larger sample sizes** are more representative of the whole ecosystem.
- **Avoiding bias** when choosing where to take samples will increase the **validity** of your results.
- Sampling, no matter how good, can only be claimed to represent a close **estimate** of an ecosystem's biodiversity.

Random and non-random sampling

The design of your sampling method will depend on the ecosystem being studied. Two approaches to assessing biodiversity are **random sampling** and **non-random sampling**.

Random sampling

You can decide the location of sampling points in an ecosystem by:

- generating random numbers, which are used as grid coordinates
- taking samples from these coordinates.

Random sampling avoids bias, but it can produce an unrepresentative picture of an ecosystem. This is especially true if the surveyed area is large. In addition, some species may be unevenly distributed and found only in certain parts of the ecosystem. Random sampling could miss these species, especially if your sample size is small because of time constraints. **Stratified sampling** can overcome this problem.

Non-random sampling

Stratified sampling

The study site is divided into smaller areas, based on the distribution of habitats. This method ensures species are not overlooked. The sampling is **more representative** of the ecosystem and **reduces sampling error**.

Worked example: Stratified sampling

Dense tree growth represents 80% of a woodland ecosystem and the remaining 20% consists of a shrubby clearing. These two areas are likely to contain different communities. If you plan to take 20 samples, 16 should be from the area with trees and four should be from the clearing. The number of samples is therefore proportional to the size of each area. Within each sub-region of your ecosystem, the location of each sample is decided randomly.

Systematic sampling

A **transect** is used where environmental gradients exist (e.g. soil pH or light intensity changes across a habitat). You can investigate whether the distribution of organisms also changes across the habitat.

Revision tip: Line or belt?

A line transect involves identifying the *presence* of species at set points along a line (usually marked with a tape). Belt transects are more detailed; the *abundance* of species can be estimated (e.g. by using quadrats).

Summary questions

1. State three factors that might limit the sample size in an ecological survey. *(3 marks)*
2. Outline how you would estimate the abundance of plant species from a rocky shoreline to the edge of sand dunes 30 metres inland. *(4 marks)*
3. Imagine you are planning to assess the biodiversity of farmland. The farmland has an area of 1 km^2, comprising 700 m^2 of grazed grassland, a shrubby area of 200 m^2, and a small orchard of 100 m^2. Outline the sampling method you would use to produce the most valid representation of this ecosystem's biodiversity. *(6 marks)*

11.3 Sampling techniques

Specification reference: 4.2.1(b)(ii) and (c)

Once a sampling strategy has been chosen, the techniques and equipment used for sampling will depend on the type of organisms being sampled.

Techniques for sampling organisms

▼ **Table 1** *Techniques and equipment for sampling animals*

Technique	Which animals are sampled?	How is it used?
Pooter	Insects	Insects are sucked into a chamber
Sweep nets	Insects in long grass	The net is swept through the habitat
Pitfall traps	Small, crawling invertebrates	A hole in the ground traps organisms
Tree beating	Tree-dwelling invertebrates	Trees are shaken and organisms fall on a sheet
Kick sampling	River-dwelling organisms	A river bed is disturbed and organisms are captured in a net

Plants are sampled using **quadrats**, which are square frames. The species present in the quadrat can be observed, identified with a key, and counted.

▼ **Table 2** *The two quadrat designs for sampling plants*

Quadrat design	How is it used?
Point quadrat	Pins are pushed through holes in a bar that spans the quadrat. All species that touch the pins are identified and recorded.
Frame quadrat	The quadrat is divided into a grid of smaller squares. Species are identified and their abundance can be estimated.

Measuring abiotic factors

Ecologists often measure abiotic (environmental) factors in the areas where they conduct species sampling.

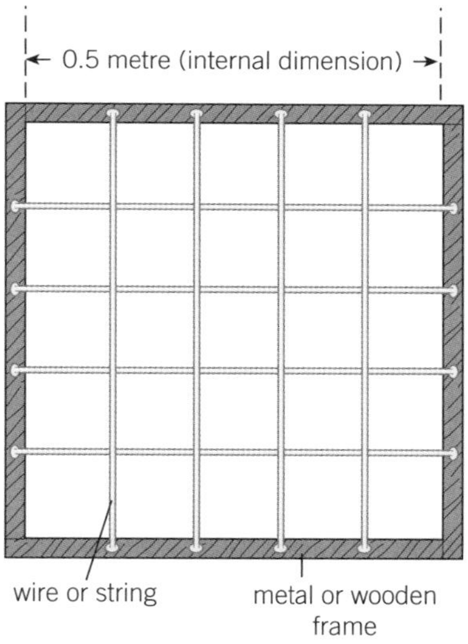

▲ **Figure 1** *A frame quadrat*

▼ **Table 3** *Examples of abiotic factors that can be measured*

Abiotic factor	Sensor	Units of measurement
Wind speed	Anemometer	$m\,s^{-1}$
Light intensity	Photometer	lux
Temperature	Thermometer	°C

Revision tip: Richness and evenness

An accurate recording of species **richness** (an area's total number of species) requires a good **identification key**. Species **evenness** is a comparison of species' numbers; this requires an accurate method for **measuring the abundance** of each species.

Quadrats can be used to record species abundance in two ways:

1 Count the *absolute* number of individuals in the quadrat (i.e. density per m^2 is recorded).

2 *Estimate* the percentage of a quadrat covered by a species.

Revision tip: Drop the 'n'

Ecological sampling is carried out with a quadrat. Remember to leave out the 'n'!

Summary questions

1 Pitfall traps are covered. Suggest why. *(1 mark)*

2 Suggest why the percentage cover of a species rather than its density might be recorded. *(2 marks)*

3 An anemometer measured a wind speed of $6\,m\,s^{-1}$. Express this speed in $km\,hr^{-1}$. Give your answer to 2 significant figures. *(2 marks)*

11.4 Calculating biodiversity

Specification reference: 4.2.1(d)

Species diversity is a measure of the number of species in a habitat (species richness) and the relative abundance of individuals in each of these species (species evenness). Both factors must be taken into account in the calculation of species diversity.

Calculating species biodiversity

Species diversity is calculated using the Simpson's Diversity Index.

Simpson's Diversity Index (D) = $1 - \left[\Sigma \left(\frac{n}{N}\right)^2\right]$

where n is the number of individuals of a particular species and N is the total number of all individuals of all species. Remember, Σ is the symbol for 'sum'. In this case, we need to calculate $\left(\frac{n}{N}\right)^2$ for each species and then add the values together.

Worked example: species diversity calculation

	Field A (adjacent to intensively farmed land)			Field B (adjacent to extensively farmed land)		
	n	$\frac{n}{N}$	$\left(\frac{n}{N}\right)^2$	n	$\frac{n}{N}$	$\left(\frac{n}{N}\right)^2$
Cocksfoot grass	60	0.60	0.3600	40	0.40	0.1600
Timothy grass	29	0.29	0.0841	20	0.20	0.0400
Meadow buttercup	4	0.04	0.0016	13	0.13	0.0169
Cowslip	1	0.01	0.0001	8	0.08	0.0064
White clover	2	0.02	0.0004	13	0.13	0.0169
Dandelion	4	0.04	0.0016	6	0.06	0.0036
Σ	–	–	0.4478	–	–	0.2438
$1-\Sigma$	–	–	0.5522	–	–	0.7562

In this example, n is the percentage cover of each plant species. N is therefore 100.

A high biodiversity is reflected by a high value of Simpson's index. Low values indicate an ecosystem is dominated by a few species and is unstable. In this example, Field B has the higher Simpson's index value and we can conclude that its biodiversity is greater than that of Field A.

Revision tip: What is represented by *n*?

We looked at methods for sampling biodiversity in the previous two topics. Remember, you are likely to measure the **percentage cover** of plant species if you sample using quadrats. If that is the case, the value for n will be a percentage rather than the absolute number of individuals.

Summary questions

1 State the difference between n and N in the Simpson's Diversity Index formula. *(2 marks)*

2 Explain why an ecosystem with a low value of D is vulnerable to environmental change. *(3 marks)*

3 Calculate the Simpson's Diversity Index for the two habitats in the following table. Which habitat is more likely to be sensitive to environmental change? *(3 marks)*

Species	Habitat A	Habitat B
Woodrush	1	4
Holly (seedlings)	11	7
Bramble	1	3
Yorkshire Fog	1	2
Sedge	2	3

11.5 Calculating genetic biodiversity

Specification reference: 4.2.1(e)

Genetic diversity determines how easily a species is able to adapt to changes in its environment. As you learned in Topic 10.8, Changing population characteristics, variation is necessary for evolution to occur. A species with high genetic diversity will contain a wide range of traits on which natural selection can act. Therefore genetic diversity increases the chance of a species adapting and surviving to environmental changes. There are several ways of measuring the genetic diversity of a species, which we discuss here.

Factors affecting genetic diversity

The genetic diversity of a population can increase (i.e. the number of alleles in the gene pool increases) due to:

- DNA mutation (see Topic 10.5, Types of variation)
- Gene flow from another population (i.e. breeding between populations of the same species).

Genetic diversity decreases due to:

- Selective breeding
- Captive breeding
- Genetic bottlenecks (i.e. when a population is reduced to a small size because of disease, habitat destruction, or migration).

Measuring genetic diversity

You read about the genetic code in Topic 3.9, DNA replication and the genetic code. A gene is a DNA sequence that codes for a polypeptide. Members of a species have the same genes, but they can have different versions of the same gene. These gene variants are known as **alleles**.

Genetic diversity is determined by the variation in a species' genes. A species that has a high percentage of genes with only one possible variant has low genetic diversity. A species that has a high percentage of genes with several possible variants has high genetic diversity.

Several methods can be used to calculate genetic diversity:

- the number of alleles per gene
- heterozygosity: the proportion of individuals in a population that have two different alleles for a particular gene
- the proportion of genes for which more than one allele exists. A gene that has two or more possible variants/alleles is known as a **polymorphic gene**. A gene for which only one variant/allele exists is called a monomorphic gene and the calculation is:

$$\text{Proportion of polymorphic genes} = \frac{\text{Number of polymorphic genes}}{\text{Total number of genes}}$$

Synoptic link

You learned about alleles in Topic 6.3, Meiosis. We looked at why genetic variation is necessary for evolution in Topic 10.8, Changing population characteristics.

Key term

Alleles: Different versions of the same gene (i.e. gene variants).

Revision tip: 'Total' means 'total sampled'

You should note that the 'total number of genes' used in the calculation tends to be a small sample of genes rather than every gene in a species' genome.

Worked example: Genetic diversity calculation

The following table shows genetic data for three populations of monkeys.

	Blue monkey population	Vervet monkey population 1	Vervet monkey population 2
Sample size (number of individuals)	93	124	364
Number of genes studied	33	23	18
Number of polymorphic genes	5	4	3
Average heterozygosity (%)	4.6	5.6	4.0

Q1 Is it possible to conclude which population has the highest genetic diversity?

Q2 What could be done to improve our confidence in the results?

A1 The proportions of polymorphic genes in each population are:

Blue monkeys: $\frac{5}{33} = 0.15$ (or 15%)

Vervet monkey population 1: $\frac{4}{23} = 0.17$ (or 17%)

Vervet monkey population 2: $\frac{3}{18} = 0.17$ (or 17%)

Based on average heterozygosity, population 1 of vervet monkeys has the highest genetic diversity. However, based on the proportion of polymorphic genes, the two vervet monkey populations have the same genetic diversity.

A2 Increasing the sample sizes and analysing a greater number of genes would improve confidence levels in the results. The genes studied represent approximately 0.1% of the total number of genes in these species. Studying more genes will increase the accuracy of the genetic diversity estimates.

Summary questions

1 Describe how genetic diversity can decrease in a natural population. *(3 marks)*

2 Explain the importance of genetic diversity to species survival. *(2 marks)*

3 Complete the following table and discuss which species has the greatest genetic diversity. *(5 marks)*

Species	Number of monomorphic genes studied	Number of polymorphic genes studied	Percentage of polymorphic genes (%)	Average heterozygosity (%)
A	28	8		4.2
B		10	22	7.0
C	8	3		5.4

11.6 Factors affecting biodiversity

Specification reference: 4.2.1(f)

Our impact on ecosystems is a more significant issue now than at any other time in history because of the continuing rise in the global human population. In this topic you will examine the direct and indirect impacts of human activities on biodiversity.

Human impacts on biodiversity

Human activity	Direct impacts on biodiversity	Indirect impacts on biodiversity
Forestry management and deforestation	Removal of native trees (e.g. deciduous woodland in the UK) has **destroyed habitats**, causing animal migrations or death. Sometimes non-native trees are grown instead to meet the demands of the timber and fuel industries, but biodiversity is lowered because fewer species are grown.	
Agriculture	**Deforestation** to clear land for farming. **Hedgerow removal** (to maximise space available for crops and machinery). **Intensive farming** uses chemicals (e.g. **pesticides and herbicides**) to kill plants and animals that are considered pests or weeds. For example, **neonicotinoid insecticides** have been blamed for reducing populations of bee species. **Monocultures** (growing a single crop) reduce the number of animals that can use the farmland as a habitat.	Fertilisers can pass into watercourses, contaminating neighbouring ecosystems (e.g. **eutrophication**). The demand for water (e.g. for irrigation of farmland) has resulted in rivers and lakes being **drained**, decimating aquatic ecosystems. For example, the Aral Sea has shrunk to 10% of its original volume due to irrigation schemes.
Fossil fuel combustion	Habitat destruction for access to fossil fuel sources	**Acid rain** production (due to air pollution) has contributed to habitat destruction (e.g. deforestation). **Climate change** (due to greenhouse gas emissions) has caused habitat destruction (e.g. desertification, melting ice in Arctic habitats, and changes in sea temperatures and currents).

Revision tip: Eutrophication

Eutrophication results from an increase in nutrients in freshwater lakes and rivers. These nutrients can enter the water from sewage or fertilisers. The nutrients encourage the growth of algae, which prevents sunlight reaching deeper aquatic plants. When these plants die, decomposers deplete oxygen in the water, which further decreases biodiversity.

Summary questions

1 State two ways in which climate change has damaged ecosystems. *(2 marks)*

2 Describe how the rising human population has affected biodiversity in aquatic ecosystems. *(4 marks)*

3 Intensive farming is characterised by the use of chemicals (e.g. fertilisers and pesticides), high costs and a large input of labour. This contrasts with extensive farming, which requires fewer chemicals, and less money and labour. Contrast the advantages and disadvantages of the two forms of farming. *(3 marks)*

11.7 Reasons for maintaining biodiversity

Specification reference: 4.2.1(g)

From a human perspective, the maintenance of biodiversity is important for a number of economic, ecological, and aesthetic reasons, which we examine here.

Why is biodiversity important?

Aesthetic reasons

Plant and animal life provides enrichment, relaxation, and inspiration; we all enjoy looking at a panda cub or strolling through beautiful woodland.

Revision tip: Sustainability

Sustainability is the concept of human populations meeting energy and food requirements without compromising biodiversity (or the ability to meet requirements in the future).

Economic reasons

- **Discovery of useful genes or compounds:** approximately half of our medicines contain a chemical derived from animal or plant species. Extinct species may have possessed genes that would have been useful in medical research or agriculture in the future.
- **Sustainable** removal of resources helps land remain economically viable for longer.
- High genetic diversity increases the chance of species (e.g. crop species) **adapting to future environmental change**. Greater biodiversity increases the range of traits available for **artificial selection** in the future.
- Money can be made through **tourism** in biodiverse areas.

Ecological reasons

As you learned in Topic 11.4, Calculating biodiversity, ecosystems with low biodiversity are unstable and especially vulnerable to environmental change. Species within a community are interdependent; the removal of one species can disrupt the rest of its food chain. For example, the extinction of a food source is likely to reduce the populations of its predators. The disappearance of insect pollinators (e.g. bee species) is liable to have a negative impact on populations of flowering plants.

Revision tip: Keystone species

A keystone is the stone that allows an arch to be self-supporting and prevents it collapsing. A keystone species prevents an ecosystem from collapsing—it has a significant effect on other species. An ecosystem is dramatically altered if a keystone species is removed.

For example, for part of the year, acorn banksia trees are the only source of nectar for honeyeater birds. Later in the year, honeyeaters pollinate other plants. The acorn banksia is a keystone species because the honeyeater population (and therefore many plant populations) would dwindle without it.

Summary questions

1 Outline two economic reasons for maintaining biodiversity. *(2 marks)*

2 Suggest two long-term benefits of sustainable forestry. *(2 marks)*

3 Sea otters eat urchins, which eat kelp (a species of seaweed). Explain why sea otters are a keystone species in kelp forest habitats. *(2 marks)*

11.8 Methods of maintaining biodiversity

Specification reference: 4.2.1(h) and (i)

Human populations can stem the rate of species extinctions by reducing levels of deforestation and intensive agriculture, and limiting the rate of global warming. However, more proactive approaches can be used to maintain or restore biodiversity. These conservation methods are our focus in this topic.

Methods for maintaining biodiversity

Scientists can either conserve species ***in situ*** (i.e. in their natural habitat) or ***ex situ*** (i.e. outside their natural habitat).

▼ **Table 1** *In situ conservation methods*

In situ method	Possible features
Wildlife reserves	Human access is restricted (e.g. poaching is banned or regulated) Animals are fed Reintroducing species Culling (i.e. removing invasive species) Preventing succession, which is a natural process of ecological change that occurs over many years (e.g. the use of controlled grazing maintains heathland ecosystems and prevents the process of succession, which would result in the formation of woodland)
Marine conservation zones	Limits placed on hunting and fishing within refuge areas

Revision tip: Conserve or preserve?

'Conservation' and 'preservation' are two terms with subtle differences. Conservation is the active and sustainable management of an ecosystem, whereas preservation leaves an ecosystem undisturbed (i.e. without human intervention).

▼ **Table 2** *Ex situ conservation methods*

Ex situ method	Possible features
Botanic gardens	Plants experience optimum conditions (e.g. ideal soil nutrients and a lack of pests)
Seed banks	Seeds from many plant species are stored in conditions that enable them to remain viable for centuries
Captive breeding programmes (e.g. in zoos or aquatic centres)	Production of offspring in human-controlled environments Reintroduction of species into natural habitats Animals experience optimum nutrition, a lack of predators, and medical attention

Conservation agreements

National and international agreements enable conservation targets and rules to be created. Important agreements include:

- The Convention on International Trade in Endangered Species (**CITES**) – regulates the trade of more than 35 000 species.
- The **Rio Convention** – targets were agreed for sustainability, reducing desertification, and limiting greenhouse gas emissions.

Summary questions

1 State the principal outcomes of the Rio Convention. *(2 marks)*

2 Suggest three advantages of using seed banks rather than botanic gardens to conserve plant species. *(3 marks)*

3 Evaluate the advantages and disadvantages of *ex situ* conservation in comparison to *in situ* methods. *(5 marks)*

Chapter 11 Practice questions

1 The genetic diversity of four populations of a species was being studied. Data for the four populations are shown below.

Population	Number of loci studied	Number of monomorphic genes
A	45	37
B	42	38
C	20	16
D	33	22

a Based on the data in the table, which population has the highest genetic diversity? Show your working and explain your conclusion. *(3 marks)*

b Comment on the reliability of conclusions drawn from this data. *(2 marks)*

2 More than 95% of the species that have existed on Earth during the past 3.5 billion years are now extinct.

a In 2014 scientists calculated the relative threat posed to biodiversity by different human activities. 37% of the total threat was estimated to be from human exploitation of ecosystems (e.g. fishing and hunting).

State two other threats to biodiversity posed by human activity. *(2 marks)*

b The highest estimate for the extinction rate of species is 0.7% of existing species per year.

Assuming an extinction rate of 0.7% per year and a current global species count of 5 million species, how many species will exist after
i 1 year *(1 mark)* **ii** 3 years *(1 mark)*

Give your answers to 4 significant figures.

c The International Union for Conservation of Nature (IUCN) decide which species are considered to be threatened with extinction. For example, 26% of mammals were placed on the IUCN's Red List of Threatened Species in 2014.

Describe how sampling techniques and data analysis can be used to monitor the population size of a mammalian species to decide whether it should be placed on the Red List of Threatened Species. *(6 marks)*

3 Which of the following statements best describes stratified sampling?

A Selecting sampling locations at random.

B Sampling based on availability.

C Sampling sub-groups in proportion to their relative sizes.

D The use of a transect for sampling. *(1 mark)*

4 Which of the following sampling techniques would be most appropriate to use when sampling river-dwelling invertebrates?

A Kick sampling **C** Quadrats

B Pooters **D** Pitfall traps *(1 mark)*

5 Which of the following is a property of a habitat with a low value of Simpson's Diversity Index?

A A large number of species

B Few ecological niches

C Complex food webs

D Environmental change produces small effects *(1 mark)*

12.1 Animal and plant pathogens

Specification reference: 4.1.1(a)

This chapter focuses on communicable diseases, which are caused by infective microorganisms known as **pathogens**.

Types of pathogen

Pathogens can be categorised into four groups: bacteria, fungi, protoctista, and viruses.

Type of pathogen	Mode of action	Appearance	Examples of diseases
Bacteria	Disease symptoms are often caused by **toxin production**	**Prokaryotic** cells (see Topic 2.6, Prokaryotic and eukaryotic cells) Shapes include rod (bacilli), spherical (cocci) and spiral	Tuberculosis (TB) Bacterial meningitis Ring rot
Fungi	They **secrete enzymes** that digest living cells, enabling the fungus to spread through tissue	**Eukaryotic** organisms (see Topic 2.4, Eukaryotic cell structure)	Ring worm Black sigatoka
Protoctista	They often consume the cell material of their host	**Eukaryotic** cells	Malaria Potato blight
Viruses	They **insert genetic material** into their host's DNA, taking control of cell metabolism	Usually considered **non-living** Protein coat enclosing genetic material	Influenza Tobacco mosaic virus

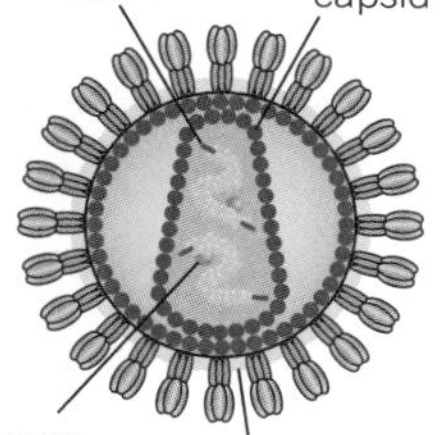

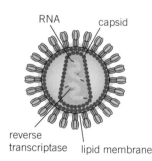

▲ **Figure 1** *The structure of HIV (Human Immunodeficiency Virus)*

Summary questions

1 State two eukaryotic kingdoms that contain pathogenic species. *(2 marks)*

2 Describe the typical cause of symptoms for diseases resulting from **a** bacterial infection **b** fungal infection. *(2 marks)*

3 Evaluate whether viruses should be considered organisms. *(3 marks)*

Key terms

Communicable disease: A disease caused by a pathogen, which can be transmitted to another organism.

Pathogen: A disease-causing organism.

Revision tip: Are you alive?

Viruses lack many traits that define living organisms (e.g. they cannot grow, synthesise proteins, or reproduce independently, and they lack membranes). They lead a 'borrowed life', relying on host cells to reproduce, and exploiting their metabolism. Some scientists think viruses occupy a grey area between living and non-living.

Revision tip: We're friendly, really ...

Most species of bacteria, protoctista, and fungi do not cause disease–only a small number of species are pathogenic.

Revision tip: Please, use my full title ...

Try to use the full binomial name of pathogens (e.g. *Mycobacterium tuberculosis*). Abbreviations are acceptable once the full name has been used once.

12.2 Animal and plant diseases

Specification reference: 4.1.1(a)

You were introduced to the four pathogenic taxa in Topic 12.1, Animal and plant pathogens. Now you will learn about some of the diseases that can be caused by these pathogens.

Plant diseases

▼ **Table 1** *Diseases of plant species*

Disease	Pathogen	Symptoms
Potato blight	*Phytophthora infestans* (a protoctist)	Hyphae (branching structures) penetrate cells, destroying tubers, leaves, and fruit
Ring rot	*Clavibacter michiganensis* (a bacterium)	Destroys vascular tissue in leaves and tubers
Tobacco mosaic virus	TMV (a virus)	Mosaic patterns of discoloration on leaves, flowers, and fruit
Black sigatoka	*Mycosphaerella fijiensis* (a fungus)	Hyphae penetrate and digest leaf cells, turning leaves black

Animal diseases

▼ **Table 2** *Diseases of animal species*

Disease	Pathogen	Symptoms
Malaria	*Plasmodium spp.* (protoctists)	Infects erythrocytes and liver cells, causing fever and fatigue
Tuberculosis (TB)	*Mycobacterium tuberculosis* (a bacterium)	Destroys lung tissue, resulting in coughing, fatigue, and chest pain
HIV/AIDS	Human immunodeficiency virus (HIV)	Infects T helper cells (see Topic 12.6, The specific immune system), thereby inhibiting the immune system
Athlete's foot	*Tinia pedia* (a fungus)	Digests skin on people's feet, causing cracking and itchiness

Revision tip: Retroviruses

HIV is a retrovirus, which means it contains RNA rather than DNA. It also contains an enzyme called reverse transcriptase, which produces a DNA copy of its RNA genome. The viral DNA is incorporated into the DNA of a T helper cell. Copies of the virus can then be produced.

Go further: Ebola

Ebola is a disease caused by a virus. It causes internal bleeding and kills approximately 50% of those infected, on average. The disease was first described in human populations in 1976. More than 20 000 cases were diagnosed in 2014, which represents an epidemic. The virus is communicable, but can be passed between humans only through direct contact with body fluids.

Suggest the difficulties that will be encountered when treating diseases such as Ebola, which is caused by a pathogen that has only recently evolved to infect humans.

Summary questions

1 State one similarity and one difference between potato blight and black sigatoka. *(3 marks)*

2 Describe how HIV is able to replicate. *(4 marks)*

3 A new strain of the H1N1 virus caused a pandemic (a worldwide outbreak) of influenza in 2009. Suggest why the new strain resulted in a pandemic. *(2 marks)*

12.3 The transmission of communicable diseases

Specification reference: 4.1.1(b)

You have learned that a communicable disease is one caused by pathogens, which can be transmitted between organisms. The subject of this topic is *how* these pathogens are passed on. Various modes of transmission exist, which can be categorised as direct or indirect.

Pathogen transmission between animals

▼ **Table 1** *Animal pathogens: modes of transmission*

Mode of transmission		Description	Examples
Direct	Contact	Contact with skin, or body fluids	Bacterial meningitis
	Entry through the skin	E.g. wounds, bites, or infected needles	HIV/AIDS Septicaemia
	Ingestion	Consumption of contaminated food or drink	Amoebic dysentery
Indirect	Fomites	Inanimate objects (e.g. bedding or clothes) that transfer pathogens	Athlete's foot
	Inhalation	Breathing in droplets containing pathogens	Influenza
	Vectors	Anything that carries a pathogen from one host to another is a vector (e.g. water, and many different animals)	Malaria (vector = mosquitoes)

Pathogen transmission between plants

▼ **Table 2** *Plant pathogens: modes of transmission*

Mode of transmission		Description	Examples
Direct	Contact	Contact between a healthy plant and a diseased plant	TMV Potato blight
Indirect	Soil contamination	Pathogens, or reproductive spores, move into the soil from infected plants	Black sigatoka Ring rot
	Vectors	Wind, water, and animals can act as vectors to transmit plant pathogens	*P. infestans* spores can be carried by air currents, causing blight to spread

Summary questions

1. Outline the social and economic factors that increase the risk of a communicable disease being spread. *(4 marks)*
2. Explain what is meant by a fomite, and state two diseases that are contracted through fomite contact. *(3 marks)*
3. Suggest why potatoes cannot be grown for at least two years on land that has supported plants with ring rot. *(2 marks)*

12.4 Plant defences against pathogens

Specification reference: 4.1.1(c)

Plants, as you have seen already in this chapter, can be vulnerable to infections by pathogens. In response, plants have evolved a range of defences to fend off their pathogenic attackers.

Types of defence

Physical defence

Callose is a polysaccharide formed from β-glucose monomers, joined with 1,3 glycosidic bonds (and some 1,6 linkages). It is largely linear (with a few branches) but helical. Callose is produced in response to pathogenic attacks and deposited in cell walls, plasmodesmata (i.e. pores in cell walls), and in sieve plates. The callose acts as a barrier to prevent further infection.

Chemical defence

Plants have evolved a range of chemical defences against pathogens and pests.

Revision tip: Recognition and response

Receptors in plant cells recognise molecules on (or produced by) pathogens. This triggers a cascade of reactions, which switches on genes to produce defensive chemicals and molecules such as callose.

▼ **Table 1** *Plant chemical defences against pests and pathogens*

Type of chemical defence	Examples
Insect repellents	Citronella, produced by lemon grass
Insecticides	Pyrethrins, produced by chrysanthemums
Antibacterial compounds	Gossypol, produced by cotton
Antifungal compounds	Saponins, produced by many species (e.g. soapworts)
Anti-oomycetes	Glucanase enzymes, which destroy cell walls in *P. infestans*
General toxins	Cyanide compounds

Summary questions

1 State two types of chemical defence produced by plants against pathogens. *(2 marks)*

2 Suggest why callose is deposited in **a** cell walls **b** plasmodesmata **c** sieve plates during an attack by a pathogen. *(3 marks)*

3 Compare the structures of callose and cellulose. *(4 marks)*

12.5 Non-specific animal defences against pathogens

Specification reference: 4.1.1(d) and (e)

Like plants, animals have evolved a range of defences against pathogens. These defences can be divided into mechanisms to combat specific pathogens and non-specific defences, which have evolved to repel a wide range of pathogens. This topic focuses on non-specific defences against pathogens.

Primary defences

Primary defences are the barriers that prevent pathogens from entering the body. They include: the **skin**, the **conjunctiva** (membrane covering the eye), **mucus**, and **ciliated epithelia** in airways, and the mucus layer and **acidic conditions** in the stomach and vagina.

Repairing the primary defences: blood clotting

Cuts to the skin leave an organism open to infection. The evolution of a blood clotting system enables repairs to be made to primary defences whenever they are damaged.

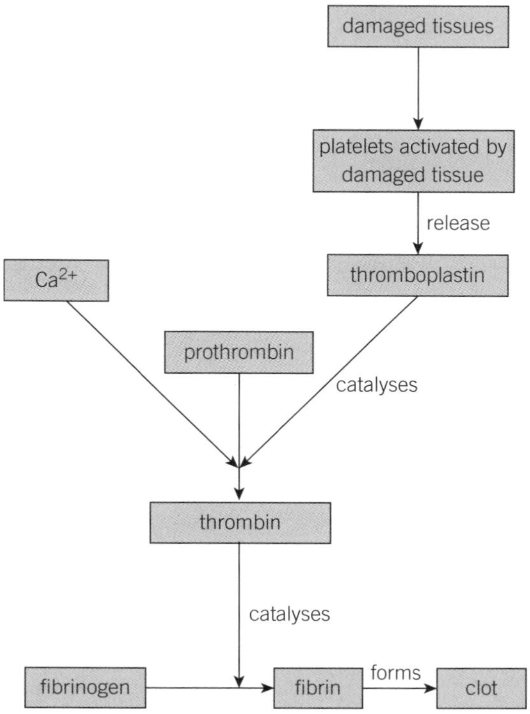

▲ **Figure 1** *The cascade of reactions that lead to blood clotting*

Secondary defences: inflammation and phagocytes

Pathogens can evade primary defences and infect an animal. Infections trigger internal non-specific responses: inflammation and phagocytosis.

▼ **Table 1** *The key features of inflammation and phagocytosis*

	What is the process?	How does it help?
Inflammation	**Mast cells** (which are leucocytes) release **histamines**, which dilate blood vessels and cause more plasma to move into tissue fluid. This **raises temperature** and causes **swelling**.	High temperature reduces the rate of pathogen reproduction. Inflammation is thought to be protective (e.g. isolating pathogens).
Phagocytosis	The phagocyte engulfs the pathogen The pathogen is enclosed in a vacuole (called a **phagosome**) **Lysosome** fuses with phagosome (forms a **phagolysosome**) **Enzymes** released by the lysosome digest the pathogen	Destruction of pathogenic cells

Synoptic link

You were introduced to the concept of phagocytosis in Topic 5.4, Active transport.

Revision tip: Helpful chemicals

Two sets of molecules assist phagocytes: cytokines and opsonins. **Cytokines** are cell-signalling molecules that, among other roles, attract phagocytes to sites of infection. **Opsonins** bind to pathogens and mark them for phagocytosis. Phagocytes have receptors that bind to opsonins.

Summary questions

1. Why are primary defences and phagocytosis known as non-specific defences? *(1 mark)*
2. Explain the roles of thromboplastin and thrombin in blood clotting. *(2 marks)*
3. Explain the role played by mast cells in defence against pathogens. *(3 marks)*

12.6 The specific immune system

Specification reference: 4.1.1(f), (g), (h), (i), and (k)

In animals, non-specific defences, such as phagocytosis, combat pathogens as soon as they infect an organism. The specific immune system targets particular pathogens but takes longer to respond.

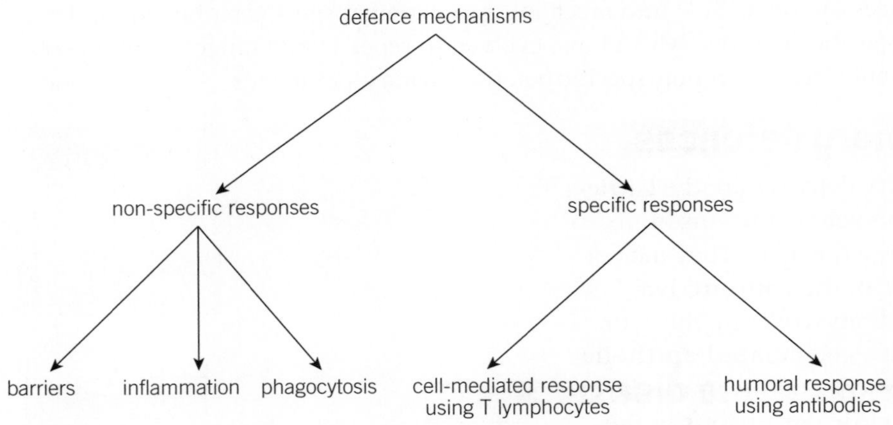

▲ **Figure 1** *An overview of animal defence mechanisms against pathogens*

Lymphocytes

Lymphocytes are white blood cells (leucocytes) that perform a variety of roles within the specific immune system.

▼ **Table 1** *Types of T and B lymphocyte*

Type of lymphocyte		Role
T lymphocyte	T helper cells	Produce **cytokines**, which stimulate B cells and other T cells
	T killer cells	Produce **perforin**, which damages the cell membranes of pathogens
	T memory cells	Recognise antigens from previous infections (**immunological memory**)
	T regulator cells	Control the immune system (**preventing autoimmune responses**)
B lymphocyte	Plasma cells	Produce **antibodies**
	B effector cells	Divide to **form plasma cell** clones
	B memory cells	Remember specific antigen (enables rapid **secondary immune response**)

Cell-mediated vs humoral immune responses

Specific immune responses are either cell-mediated or humoral.

▼ **Table 2** *The two branches of the specific immune system*

	Cell-mediated immunity	Humoral immunity
What happens?	**Antigen-presenting cells** (e.g. phagocytes) activate **T-helper cells**, which stimulate **phagocytosis**, and T memory and **killer cell** production No antibodies	**Clonal selection** of antigen-specific B cell **Clonal expansion** to produce **plasma cells** and B memory cells **Antibody production**
Typical targets	Viruses and cancerous cells	Bacteria and fungi

Revision tip: Good sense of humor?

The humoral immune response does not involve stand-up comedy; 'humor' is an old-fashioned word for body fluid, and the humoral response takes place in blood plasma and tissue fluid.

Antibodies

Antigens are molecules on cells (or viruses) that the immune system can use to detect infection. An organism's own cells have *self* antigens, whereas pathogens exhibit *non-self* antigens. **Plasma cells** in the immune system produce **antibodies** that are specific to a pathogen's non-self antigens.

The variable region is different in each antibody, which enables each antibody to bind to a specific antigen (on a pathogen). Depending on the type of antibody, a variety of fates await the pathogen.

▼ **Table 3** *Types of antibody defence*

Method of antibody defence	What happens?
Opsonisation	Antibody acts as an opsonin (speeding up phagocytosis).
Agglutination	Antigen–antibody complexes clump together. This clump is too large to enter cells and enables phagocytes to engulf several pathogens at once.
Neutralisation	Antibodies bind to toxins, rendering them harmless.

Autoimmune diseases

The immune system can malfunction and stop recognising self antigens. The body's cells are attacked by its own immune system. This is known as **autoimmune disease**.

▼ **Table 4** *Examples of autoimmune diseases*

Autoimmune disease	Body part affected	Symptoms
Grave's disease	Thyroid gland	Overactive thyroid, causing weight loss and muscle weakness
Vitiligo	Melanocytes	Loss of skin pigmentation
Type 1 diabetes	Pancreatic β-cells	Lack of insulin production; loss of blood glucose regulation

Summary questions

1 Outline the role of T killer cells. *(2 marks)*

2 Explain how agglutination limits bacterial infection of cells. *(3 marks)*

3 Suggest why someone with HIV/AIDS may not exhibit a secondary immune response despite already encountering a pathogen. *(2 marks)*

Key terms

Antigen: A molecule that triggers an immune response (i.e. antibody production).

Antibody: a glycoprotein produced in response to the presence of an antigen.

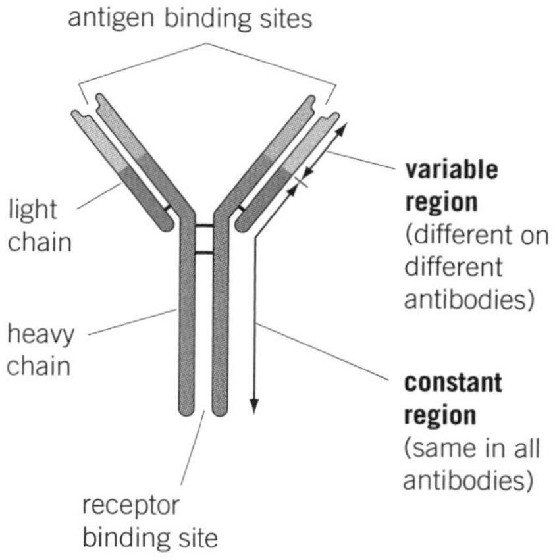

▲ **Figure 2** *The structure of an antibody (a glycoprotein comprising four polypeptide chains)*

Revision tip: Primary vs secondary response

The primary response of the immune system is the production of antibodies against a pathogen that has not been encountered before. This can take days or sometimes weeks. The secondary response occurs when the pathogen infects an organism again. The response is much quicker because memory cells retain the ability to recognise the set of antigens on the pathogen.

12.7 Preventing and treating disease

Specification reference: 4.1.1(j), (l), (m), and (n)

Revision tip: Antibiotics treat bacterial infections

Antibiotics cannot be used to treat viral infections. The metabolic reactions that are targeted by antibiotics are absent from viruses.

▼ **Table 1** *The different ways in which a person can gain immunity*

	Natural	Artificial
Active	Memory cells produced following pathogenic **infection**	Memory cells produced following a **vaccination** (see below)
Passive	Fetal immunity (**maternal antibodies** cross the placenta)	**Antibodies are injected** into a person, providing temporary immunity

An animal is immune if they can be infected with a pathogen without developing any symptoms of the disease. This immunity can develop in a number of ways, which you will read about here. This topic also examines methods of treating diseases once they have been contracted (e.g. the use of antibiotics).

Vaccinations

The principle of vaccination is to persuade the body to produce antibodies and memory cells against a particular pathogen without a person contracting the disease. Vaccination of many people in a population can prevent a disease spreading; this is called **herd immunity** and prevents **epidemics**.

▼ **Table 2** *Types of vaccine*

Type of vaccine	How does it work?	Examples of diseases	Advantages/ disadvantages
Weakened, live pathogen	Modified pathogen that is alive but not pathogenic	Mumps, polio, measles, TB	Strongest response and long-lasting immunity, but (rarely) organism may revert and become pathogenic
Dead/inactivated pathogen	Pathogen is killed but its antigens are still present	Influenza, whooping cough	Stable and safer than live vaccines, but response is weaker (boosters required)
Toxoids	Modified toxins	Tetanus, diphtheria	Safe, but may not give strong response
Subunits	Isolated antigens	HIB	Vaccines for several strains produced

Sources of medicines

Vaccines are designed to prevent pathogenic infections. When someone does contract a disease, however, medicinal treatments can be given. Medicinal drugs are often derived from natural compounds.

▼ **Table 3** *Medicines from natural sources*

Medicine	Source	Properties and uses
Quinine	*Cinchona* spp.	Antimalarial, painkilling
Aspirin	*Salix alba* (willow)	Anti-inflammatory, painkilling
Penicillin	*Penicillium* fungi	Antibiotic

Antibiotic resistance

Mutation can result in the evolution of bacteria that are resistant to antibiotics (e.g. methicillin-resistant *Staphylococcus aureus* (MRSA)). The spread of antibiotic-resistant infection can be reduced by minimising the use of antibiotics (as overuse can accelerate natural selection of resistant strains) and using good hygiene practices.

Summary questions

1. Explain why artificial passive immunity does not provide long-term immunity. *(2 marks)*
2. Explain why herd immunity reduces the spread of a disease. *(2 marks)*
3. Suggest why vaccinations against tetanus require booster injections. *(3 marks)*

Chapter 12 Practice questions

1 Which of the following can be used in a vaccine?

1 An inactivated virus

2 Antibodies specific to a bacterial pathogen

3 A protein fragment from a virus

A 1, 2, and 3 are correct

B Only 1 and 3 are correct

C Only 2 and 3 are correct

D Only 1 is correct *(1 mark)*

2 Which of the following statements best describes how herd immunity is usually achieved?

A A large proportion of a population survives a disease epidemic.

B A large proportion of a population is vaccinated.

C A large proportion of a population is given antibodies specific to a pathogen.

D A large proportion of a population receives antibodies specific to a pathogen from their mothers during gestation. *(1 mark)*

3 Which of the following statements describes the typical mechanism of bacterial pathogenicity?

A Integration of genetic material into host cell chromosomes

B Enzyme secretion

C Production of toxins

D Activation of host cell genes *(1 mark)*

4 A nation has a population of 3.4×10^7 people.

In one year, this nation had 2.0×10^5 cases of tuberculosis (TB), of which 40 000 were new cases.

a What was the incidence rate (per 100 000) of TB in this nation? *(1 mark)*

b The same nation had 6.8×10^3 deaths from TB over a one-year period.

What was the mortality rate (per 100 000) from TB in this nation? *(1 mark)*

5 Suggest why attempts to develop a vaccine against HIV have been unsuccessful. *(2 marks)*

6 **a** Outline methods farmers can use to reduce the spread of disease between crop plants. *(3 marks)*

b Outline methods that hospital staff can use to reduce the spread of disease within hospitals. *(2 marks)*

7 Complete the following passage about plant responses to pathogens. Choose the most appropriate word to place in each gap. *(5 marks)*

Pathogenic molecules can bind to .... in the cell membranes of plant cells. This triggers a cascade of reactions that results in genes in the .... being switched on. A polysaccharide called .... is produced, which is deposited in cell walls and .... sieve plates. These physical barriers are further strengthened by ....

Answers to practice questions

Chapter 2

1 D *[1]*, **2** A *[1]*, **3** A *[1]*

4

Structure	Membrane-bound	Contains DNA	
Mitochondrion	YES	YES	*[1]*
Chloroplast	YES	YES	*[1]*
Lysosome	YES	NO	*[1]*
Microtubules	NO	NO	*[1]*

5 Fixing/fixation; *[1]*, Differential; *[1]*, Crystal violet; *[1]*, Cell walls *[1]*

6 9×10^{-5} m
(1 mark awarded for 90 μm) *[2]*

7 Both consist of microtubules; *[1]*
Eukaryotic flagella have a 9 + 2 arrangement of microtubules; *[1]*
Enable cell movement in both types of cell; *[1]*
Thinner in prokaryotes *[1 (max 3)]*

Chapter 3

1 B *[1]*, **2** D *[1]*, **3** A *[1]*, **4** B *[1]*

5 red; *[1]*, cuvette; *[1]*, Benedict's; *[1]*, precipitate *[1]*

6 a Phosphate (PO_4^{3-}) ions; *[1]*, From soil; *[1]*, Diffusion into roots *[1]*

b **Level 3 (5–6 marks):** Discussion of at least three of the following examples: phospholipids/cell membranes, nucleic acids, ATP, NADP, bones. Structure has been linked to function.

Level 2 (3–4 marks): Discussion of at least two of the following examples: phospholipids/cell membranes, nucleic acids, ATP, NADP, bones. Both structure and function of the relevant molecules have been discussed to some extent.

Level 1 (1–2 marks): Discussion of at least one of the following examples: phospholipids/cell membranes, nucleic acids, ATP, NADP, bones. The information is supported by limited evidence.

7 a Sweating / panting; *[1]*
Water has a high heat of vaporisation; *[1]*
A relatively large amount of energy is required to change water from liquid to gas; *[1]*
The evaporation of water from the surface of an organism has a cooling effect *[1]*

b Dipoles shown (i.e. partial positive charge on H atoms and partial negative charge on O atoms within OH groups); *[1]*
Hydrogen bond drawn between H in water and O in ethanol (or O in ethanol and H in water) *[1]*

8 a Detergent: breaks down cell membranes; *[1]*
Salt: breaks hydrogen bonds between water and DNA; *[1]*
Protease: breaks down proteins (e.g. histones); *[1]*
Alcohol: precipitates DNA *[1]*

b i purine, pyrimidine, purine (A,C,G); *[1]*
pyrimidine, pyrimidine, purine (T,T,A) *[1]*

ii two *[1]*

iii UGC; *[1]*
AAU *[1]*

Chapter 4

1 D *[1]*, **2** C *[1]*, **3** A *[1]*, **4** C *[1]*

5 a 7 *[1]*

b Place % in the column heading rather than in every box; *[1]*
Write pH values to the same number of decimal places *[1]*

c Test a greater range of pH values *[1]*

6 Complementary; *[1]*
Active site; *[1]*
Similar *[1]*

7 pH; *[1]*, Temperature; *[1]*, Enzyme concentration; *[1]*, The overall volume of solution *[1 (max 3)]*

Chapter 5

1 C *[1]*, **2** A *[1]*, **3** C *[1]*, **4** B *[1]*

5 *Cell communication*
Glycoproteins; *[1]*
Embedded/AW in the membrane bilayer; *[1]*
Act as receptors *[1]*
Exchange of substances
Partially permeable; *[1]*
Membranes surround organelles; *[1]*
Carrier/channel proteins; *[1]*
Only specific substances are allowed to pass through a membrane *[1 (max 5)]*

6 Membranes are fluid because components (phospholipids, cholesterol, and proteins) can move within the bilayer; *[1]*
Membranes are compared to mosaics because proteins are scattered throughout the bilayer, like the tiles in a mosaic *[1]*

7 Highest/er; *[1]*, Lowest/er; *[1]*
Turgor; *[1]*, Plasmolysis *[1]*

8 Water diffuses from cell B to A and C; *[1]*
Water diffuses from cell A to C; *[1]*
Water moves from high to low water potential *[1]*

Chapter 6

1 B *[1]*, **2** B *[1]*, **3** C *[1]*

4 **Level 3 (5–6 marks):** Exemplification of stem cell therapy, including discussion of different stem cell sources (e.g. multipotent stem cells, pluripotent ESCs, induced PSCs) and detailed discussion of problems arising from research and clinical trials. There is a well-developed line of reasoning which is clear and logically structured. The information presented is relevant and substantiated.

Level 2 (3–4 marks): Exemplification of stem cell therapy (e.g. use of multipotent stem cells to treat leukaemia) and some of the problems arising from research and trials. There is a line of reasoning presented with some structure. The information presented is in the most part relevant and supported by some evidence.

Level 1 (1–2 marks): Simple comments about stem cell use with little exemplification or detail. The information is basic and communicated in an unstructured way. The information is supported by limited evidence.

5 Chromatin; *[1]*
Nucleolus; *[1]*
Centrioles; *[1]*
Microtubules *[1]*

6 A = Palisade cell (Z); *[1]*
Explanation: greater rate of carbon dioxide diffusion; *[1]*
B = Sperm cell (Y); *[1]*
Explanation: the acrosome contains enzymes that enable the sperm cell to enter a secondary oocyte; *[1]*
C = Erythrocyte (X); *[1]*
Explanation: more space for haemoglobin to be stored; *[1]*
D = Neutrophil (W); *[1]*
Explanation: lysosomes contain enzymes that digest pathogens *[1]*

Chapter 7

1 A *[1]*, **2** C *[1]*, **3** C *[1]*, **4** A *[1]*

5 **a** Radius = 0.5 μm; *[1]*
Surface area = 3.14 μm^2; *[1]*
Volume = 0.52 μm^3; *[1]*
Ratio = 6:1 *[1]*

b Large surface area: volume ratio; *[1]*
Exchange can occur by diffusion alone *[1]*

6 External; *[1]*, Flattens; *[1]*,
Thorax/chest cavity; *[1]*, Pressure *[1]*

Chapter 8

1 D *[1]*, **2** A *[1]*, **3** B *[1]*, **4** D *[1]*, **5** C *[1]*

6 Plasma; *[1]*, Hydrostatic; *[1]*,
Potential; *[1]*, Venous *[1]*

7 **a** X = endothelium; *[1]*
Y = (smooth) muscle *[1]*
b Flexibility; *[1]*
Enabling stretch and recoil *[1]*

Chapter 9

1 C *[1]*, **2** D *[1]*, **3** B *[1]*, **4** C *[1]*

5 Root pressure drops when oxygen levels decrease; *[1]*
Temperature and root pressure are correlated; *[1]*
Respiratory poisons (e.g. cyanide) reduce/stop root pressure; *[1]*
Xylem sap *[1 (max 3)]*

6 W = cortex *[1]*, X = phloem *[1]*,
Y = xylem *[1]*, Z = parenchyma *[1]*

7 Mesophyll; *[1]*, Xylem vessels; *[1]*,
Cohesion; *[1]*, Capillary *[1]*

Chapter 10

1 C *[1]*, **2** D *[1]*, **3** B *[1]*

4 Mutation; *[1]*, Alleles; *[1]*,
Change in selection pressure; *[1]*, Adapted *[1]*

5 **a** Normal *[1]*
b **i** Continuous *[1]*
ii Polygenic/influenced by many genes; *[1]*
Environmental influence *[1]*
c Mean, mode, and median are the same; *[1]*
164 cm *[1]*

6 ***Level 3 (5–6 marks)***: Detailed and accurate descriptions of the types of evidence used in classification, with examples to illustrate each type of evidence. There is a well-developed line of reasoning which is clear and logically structured. The information presented is relevant and substantiated.

Level 2 (3–4 marks): Includes some accurate descriptions of the types of evidence used in classification, with examples to illustrate most types of evidence. There is a line of reasoning presented with some structure. The information presented is in the most part relevant and supported by some evidence

Level 1 (1–2 marks): Simple comments about the evidence used in classification, with few correct examples. The information is basic and communicated in an unstructured way. The information is supported by limited evidence.

Relevant scientific points include:

Biochemical/molecular evidence (including DNA and amino acids);

Some candidates may include details of the techniques used to analyse these molecules;

Examples include: cytochrome c differences between species; comparisons of hominid DNA in phylogeny.

Anatomical evidence;

Candidates may discuss limitations of anatomical evidence, including convergent evolution;

Immunological evidence;

Behavioural evidence; examples may include similarities in primate behaviour;

Embryological evidence; examples include similarities in vertebrate embryos

Chapter 11

1 **a** D; *[1]*

(11/33) × 100 = 33.3% of genes are polymorphic; *[1]*

Population D has the highest proportion of polymorphic genes *[1]*

b (Reliability is doubtful because) only a small proportion of the genome has been sampled; *[1]*

Different numbers of genes from each population have been analysed *[1]*

2 **a** Habitat, destruction/degradation/change/loss; *[1]*

Climate change; *[1]*

(Introduction of) invasive species; *[1]*

Pollution *[1 (max 2)]*

b **i** 4 965 000 *[1]*, **ii** 4 896 000 *[1]*

c **Level 3 (5–6 marks)**: Detailed description of potential sampling methodology and suggestions for approaches to analysis of the data. There is a well-developed line of reasoning which is clear and logically structured. The information presented is relevant and substantiated.

Level 2 (3–4 marks): Generally accurate description of potential sampling methodology and suggestions for approaches to analysis of the data. There is a line of reasoning presented with some structure. The information presented is in the most part relevant and supported by some evidence.

Level 1 (1–2 marks): Simple comments about either sampling methodology or analysis. The information is basic and communicated in an unstructured way. The information is supported by limited evidence.

Relevant scientific points include:

Sampling

Larger sampling area / number of samples improves validity;

Avoiding bias when sampling / random sampling;

Description of stratified sampling;

Suitable (repeatable) method for counting a mammalian species;

Regular sampling over time (several years) to monitor trends

Analysis

Use of statistical tests to analyse trends over time;

Analysis of genetic diversity within the species (e.g. heterozygosity or proportion of polymorphic alleles);

Estimation of total population size from samples;

Calculation of rate of decline (of population);

Calculation of range / distribution of species

3 C *[1]*, **4** A *[1]*, **5** B *[1]*

Chapter 12

1 B *[1]*, **2** B *[1]*, **3** C *[1]*

4 **a** 118 *[1]*, **b** 20 *[1]*

5 Danger of a live vaccine reactivating; *[1]*

HIV infects lymphocytes and evades immune detection; *[1]*

High mutation rate *[1 (max 2)]*

6 **a** Leave space between plants; *[1]*

Control / kill insect vectors; *[1]*

Rotate crops (to limit contamination via soil); *[1]*

Follow hygiene practices *[1 (max 3)]*

b Regular hand washing; *[1]*

Efficient waste disposal; *[1]*

Washing of bedding (to reduce fomite transmission) *[1 (max 2)]*

7 Receptors; *[1]*, Nucleus; *[1]*, Callose; *[1]*, Phloem; *[1]*, Lignin *[1]*

Answers

2.1

1 To enable light to pass through from below *[1]*

2 Objective magnifies the image; *[1]*

Eyepiece increases the magnification further *[1]*

3 Differential staining (described); *[1]*

Xylem will stain yellow; *[1]*

Non-lignified cell walls will stain blue *[1]*

2.2

1 *Magnification*: size of image divided by actual size of object; *[1]*

Resolution: the shortest distance between two objects that can still be distinguished as separate structures *[1]*

2 1.5; *[1]*

μm *[1]*

(or 0.0015 mm)

3 0.0112 m (or 1.12 cm); *[1]*

1.12×10^{-2} m (in standard form); *[1]*

1.1×10^{-2} m (to 2 significant figures)

[1 (the correct answer in standard form receives all three marks without working)]

2.3

1 Electron microscopes have greater resolution; *[1]*

Because the wavelengths of electrons are shorter than the wavelengths of light *[1]*

2 0.30 m (or 300 000 µm); *[1]*

3.0×10^{-1} m

[1 (the correct answer in standard form receives both marks without working)]

3 *Electron microscopes*:

Greater resolution; *[1]*

Higher magnification; *[1]*

Specimen must be dead (whereas CLSMs can use living tissue); *[1]*

Both:

3D images (but not with TEMs); *[1]*

Expensive/artefacts can be introduced in preparation *[1]*

(accept reverse arguments)

2.4

1 Both contain microfilaments (in a 9 + 2 arrangement); *[1]*

Flagella are longer than cilia *[1]*

2 Synthesis at rough ER/ribosomes; *[1]*

Transported to Golgi in vesicle; *[1]*

Fusion of vesicle with Golgi membrane; *[1]*

Modified within Golgi; *[1]*

Packaged into secretory vesicle; *[1]*

Fusion of vesicle within plasma cell membrane *[1 (max 5)]*

3 Large networks help to control movements (e.g. of organelles) within cells; *[1]*

Rapid assembly and disassembly can produce movement of whole cells (i.e. controls cell migration); *[1]*

(enables rapid) cytokinesis/cell division *[1]*

2.5

1 Maintaining turgor *[1]*

2 Other groups of organisms contain cell walls and vacuoles (other than some algal species, only plants contain chloroplasts); *[1]*

The cell walls and vacuoles in other groups, however, exhibit structural differences to plant cells (e.g. bacterial cell walls contain peptidoglycan); *[1]*

Not all plant cells contain chloroplasts *[1 (max 2)]*

3 Chloroplasts may have evolved from bacteria; *[1]*

Small bacteria may have entered larger cells (i.e. symbiosis); *[1]*

Ribosomes in chloroplasts have evolved a different structure to prokaryotic ribosomes, suggesting protein synthesis works differently in chloroplasts and bacteria *[1]*

2.6

Go further

1 The artefacts were probably caused by damage to the plasma membrane during the preparation of the cells for use in electron microscopes

2 Aerobic respiration does occur on the plasma membrane of bacteria, but without infoldings

Summary questions

1 Prokaryotic walls are made of peptidoglycan/murein; *[1]*

Eukaryotic walls are made of chitin/cellulose *[1]*

2 Increases genetic variation; *[1]*

Bacteria reproduce asexually, which limits variation *[1]*

3 *Eukaryotic*

Linear chromosomes; *[1]*

Histone proteins; *[1]*

Within a nucleus; *[1]*

Also found in mitochondria/chloroplasts; *[1]*

Prokaryotic

Naked/no nucleus; *[1]*

Circular; *[1]*

Also found in plasmids *[1 (max 4)]*

3.1

1 Carbon; *[1]*

Oxygen; *[1]*

Hydrogen *[1]*

2 *Compound*: more than one element bonded together; *[1]*

Molecule: more than one atom; *[1]*

covalently bonded *[1]*

3 **a** Cl^-, HCO_3^-, NH_4^+, Ca^{2+}, OH^- *[1]*

b CH_3COO^- *[1]*

c HCO_3^-, NH_4^-, CH_3COO^-, OH^- *[1]*

3.2

1 Oxygen attracts electrons more than hydrogen; *[1]*

Oxygen has a negative dipole/partial charge; *[1]*

Hydrogen has a positive dipole/partial charge; *[1]*

2 Hydrogen bonds; *[1]*

Water molecules have cohesion and flow together; *[1]*

Water is a good solvent/dissolves and transports important biological molecules and ions *[1]*

3 The positive charges are attracted to the δ^- O atom in water; *[1]*

The negative charges are attracted to the δ^+ H atom in water *[1]*

3.3

1 Condensation reaction; *[1]*

Between glucose and fructose; *[1]*

Glycosidic bond formed *[1]*

2 Both contain C, O, and H only; *[1]*

Both have ring structures; *[1]*

Glucose has 6 C atoms/ribose has 5 C atoms *[1]*

3 Cellulose is made from β-glucose monomers; *[1]*

Glycogen is made from α-glucose monomers; *[1]*

Cellulose has hydrogen bonds/cross-links between chains; *[1]*

Cross-links increase strength, which is required in cellulose's support role in cell walls; *[1]*

Glycogen has 1,6 glycosidic bonds, which enables branching; *[1]*

Branching makes glycogen compact; *[1]*

Branching increases the number of points at which glucose monomers can be hydrolysed; *[1]*

Both polymers are insoluble *[1 (max 6)]*

3.4

1 Colorimetry (following the Benedict's test); *[1]*

Biosensor measurements; *[1]*

Reagent strips *[1]*

2 **a** *absorption*: glucose concentration on the x-axis and absorption on the y-axis; *[1]*

Absorption decreases as glucose concentration increases; *[1]*

Explanation: more Benedict's reagent reacts as glucose concentration increases; *[1]*

Less light is absorbed (and more passes through) when less Benedict's reagent (which absorbs red light) remains; *[1]*

b *transmission*: glucose concentration on the x-axis and transmission on the y-axis; *[1]*

Transmission increases as glucose concentration increases *[1]*

3 Benedict's reagent is blue and therefore absorbs red light; *[1]*

Measurements of transmission will be more precise if only red light is allowed to reach the solution *[1]*

3.5

1 White/creamy; *[1]*

emulsion *[1]*

2 Hydrophilic phosphate; *[1]*

Faces/interacts with water (in tissue fluid and cytosol); *[1]*

Hydrophobic fatty acids; *[1]*

Form a partially permeable barrier *[1]*

3 *Similarities*: condensation reactions; *[1]*

Water produced; *[1]*

Enzymes catalyse the reactions; *[1]*

Differences: polypeptides are polymers, but triglycerides are not; *[1]*

Peptide bonds formed in polypeptides, but ester bonds formed in triglycerides; *[1]*

Triglycerides always consist of glycerol and three fatty acids, whereas polypeptides vary in the number of amino acids they contain *[1 (max 4)]*

3.6

Go further

1 They are hydrophobic

2 More intermolecular bonds form between glutenins and gliadins

3 The salt ions interfere/block charges on proteins that might otherwise cause repulsion between polypeptides

Summary questions

1 Central C atom; *[1]*

Four groups bonded to the central carbon: COOH, NH_2, H, and CH_3 *[1]*

2 **a** 0.83 *[1]*

b 3.55 cm *[1]*

3 Prepare a range of protein solutions of known concentration; *[1]*

Add a specific volume of biuret solution to each protein solution; *[1]*

Use colorimetry to plot a calibration curve (protein concentration against absorption; *[1]*

Test the unknown sample and calculate protein concentration from the calibration curve *[1]*

3.7

1 Golgi apparatus *[1]*

2 Globular proteins often need to be transported (e.g. in blood) and therefore need to be water-soluble; *[1]*

Fibrous proteins tend to have structural roles and should not dissolve in water. *[1]*

3 Keratin's flexibility increases when the number of disulfide bonds decreases; *[1]*

Hair is more flexible than nails *[1]*

3.8

1

Monomer	Polymer	Bond formed in condensation reaction
Monosaccharide	Polysaccharide	Glycosidic
Amino acid	Polypeptide	Peptide
Nucleotide	Polynucleotide	Phosphodiester

[1 mark per correct row]

2 Complementary base pairing; *[1]*

A pairs with T, C pairs with G *[1]*

3 Molecular structures are not complementary; *[1]*

Hydrogen bonds cannot form *[1]*

3.9

1 CGG; *[1]*

TTT; *[1]*

AGA *[1]*

2 Helicase; *[1]*

For unwinding and unzipping DNA double helix; *[1]*

DNA polymerase; *[1]*

For joining sugar-phosphate backbone/nucleotides within a new strand *[1]*

3 Degenerate code; *[1]*

The same amino acid might still be coded for even if one base changes within a codon (i.e. a substitution/ point mutation) *[1]*

3.10

1 Antisense *[1]*

2 DNA can be replicated; *[1]*

Increases stability/reduces damage; *[1]*

Reduces the possibility of unwanted bonding within a single strand *[1 (max 2)]*

3

	DNA replication	Transcription	Translation
Hydrogen bonds between complementary DNA base pairs are broken	Yes	Yes	No
Free DNA nucleotides are activated	Yes	Yes	No
To which base does adenine bond?	T	U	U
Phosphodiester bonds are formed	Yes	Yes	No
Peptide bonds are formed	No	No	Yes
Location	Nucleus	Nucleus	Ribosomes (rough ER)
Product	DNA	mRNA	Polypeptide

[1 mark per correct row]

3.11

1 Both contain ribose; *[1]*

Both contain phosphates and a base; *[1]*

ATP contains three phosphates/RNA nucleotides contain one phosphate; *[1]*

The base in ATP is always adenine/RNA nucleotides can contain A, U, C, or G *[1]*

2 ATP is unstable; *[1]*

Because of phosphate bonds that are easily broken *[1]*

3 ATP is found in all living cells; *[1]*

Energy is 'carried' in chemical bonds *[1]*

4.1

1 (all enzymes are) proteins/polymers of amino acids *[1]*; (all enzymes have a) globular structure *[1]*; folded *[1]*; contain active sites *[1] (max 3)*

2 a anabolic because a larger molecule (glycogen) is formed from smaller molecules (glucose) *[1]*;

b catabolic because smaller molecules (maltose) are formed from a larger molecule (amylose) *[1]*;

c catabolic because a smaller molecule (glycerol) is formed from a larger molecule (a triglyceride) *[1]*

3 alternative reaction pathway/AW *[1]*; temporary bonds between substrate and active site *[1]*; substrate bonds are strained/stretched *[1]*; less energy required to break the bonds in the substrate *[1]*

4.2

1 Denaturation *[1]*; secondary/tertiary structure of the enzyme is changed *[1]*; active site is altered *[1]*; active site is no longer a complementary shape to the substrate *[1]*; high temperatures cause weak bonds in the enzyme to vibrate and break *[1]*; pH changes result in hydrogen and ionic bonds breaking *[1] (max 5)*

2 The pH in the human body will be lower than the pH in the lakes *[1]*; enzymes in the microorganisms will not function at an optimal rate/may be denatured *[1]*

3 Psychrophile enzyme structures are more flexible *[1]*; different bonds in the tertiary structure *[1]*; fewer hydrogen bonds/more disulfide bonds *[1] (max 2)*

4.3

1 Competitive inhibitors bind to active sites/non-competitive inhibitors bind to allosteric sites *[1]*; non-competitive inhibitors alter the tertiary structure of enzymes *[1]*

2 Competitive inhibition *[1]*; similar shape/structure to the substrate/PABA *[1]*; complementary shape to the enzyme's active site *[1]*; competes with PABA for the active site *[1]*; fewer enzyme-substrate complexes formed *[1] (max 4)*

3 Negative feedback occurs when a change in a variable leads to the reversal of the change *[1]*; (in end-product inhibition) as the concentration of product increases, the reaction that produces the product is inhibited *[1]*;

4.4

1 Prosthetic groups bind permanently/tightly (whereas coenzymes bind temporarily/loosely) *[1]*; all coenzymes are organic (whereas prosthetic groups can be organic or inorganic) *[1]*

2 Otherwise digestive enzymes would cause damage to cells *[1]*; the enzymes are activated in the digestive tract *[1]*

3 Translated polypeptide *[1]* moves in vesicle *[1]* to Golgi apparatus *[1]* where prosthetic group is added *[1]*

5.1

Go further

1 Cholesterol increases the range of temperatures over which a membrane can maintain an appropriate fluidity (i.e. a fluidity that enables it to function). Cholesterol resists changes to fluidity, thereby stabilising a membrane

2 Cholesterol will maintain fluidity (and therefore permeability) at cold temperatures and enable diffusion of important particles (e.g. O_2) into the bacterial cells

Summary questions

1 Compartmentalisation of organelles *[1]*; anchors/supports cytoskeleton *[1]*; a site for chemical reactions (e.g. thylakoid membranes in chloroplasts) *[1]*; vesicle formation *[1] (max 3)*

2 a carrier proteins *[1]* because they enable facilitated diffusion of large molecules *[1]*

b glycoproteins *[1]* because the carbohydrate attached to the protein acts as a receptor *[1]*

3 Proteins could move within the membrane *[1]* and end up in the wrong positions within the membrane *[1]*

5.2

1 Increases in temperature raise membrane fluidity and therefore permeability *[1]* by increasing the kinetic energy of phospholipids *[1]*, which widens the spaces between the phospholipid molecules *[1]*

2 Proteins in the membrane are denatured *[1]*, preventing cell signalling/transmembrane transport *[1]*

3 Ethanol is lipid soluble *[1]* and can insert itself between phospholipids *[1]*, increasing membrane permeability *[1]*; extra particles can leave and enter cells *[1]*, thereby altering particle concentrations and cell content *[1]*

5.3

1 a Microvilli increase surface area *[1]*, which increases diffusion rate *[1]*;

b For facilitated diffusion of glucose *[1]*, which is large and polar and therefore cannot diffuse through the phospholipid bilayer *[1]*

2 a Diffusion though the bilayer *[1]* because CO_2 is small and non-polar *[1]*;

b Facilitated diffusion *[1]* because potassium ions are charged and therefore not lipid soluble *[1]*

3 Surface area *[1]* × concentration gradient *[1]*/ membrane thickness *[1]*

5.4

1 Molecules being transported out of cells *[1]*, which are large/polar *[1]*; often proteins *[1]*; examples: hormones/enzymes/neurotransmitters *[1]*

2 One mark per correct row

	Active transport	Facilitated diffusion
Uses carrier proteins	Yes	Yes
Particles move down concentration gradients	No	Yes
Particles move against concentration gradients	Yes	No
Requires ATP	Yes	No
At least two binding sites must be present on carrier proteins	Yes (one for the particle, one for ATP)	No

3 Larger vesicles for phagocytosis *[1]*; phagocytosis transports large solid material into cells *[1]*; pinocytosis transports small, dissolved particles and small amounts of water into cells *[1]*

5.5

1 a False *[1]*

b True *[1]*

c False *[1]*

2 a increase in cell size (possible lysis/bursting, but this is unlikely because the difference in water potential is small) *[1]*

b No change *[1]*

c The cell shrinks/experiences crenation *[1]*

3 Water potential in the soil water is reduced *[1]*; water leaves plant (root hair) cells via osmosis *[1]*; plant cells plasmolyse *[1]* and lose turgor *[1]*

6.1

Go further

1 9.6 hours

2 Possible health problems include DNA damage, cancer, or malnourishment, such as a shortage of certain nutrients

Summary questions

1 a Organelles *[1]* (named) biochemicals *[1]*

b DNA *[1]*

2 The timing of each step in the cell cycle is crucial *[1]*; checkpoints regulate the sequence of events in the cycle *[1]*; the cycle will be halted if errors are detected *[1]*

3 a Although cells are not dividing in interphase, they are preparing to divide (by synthesising molecules and organelles) *[1]*

b Cells in G_0/outside the cell cycle/some differentiated cells *[1]* although these cells are not 'resting' (i.e. they are metabolically active despite not dividing) *[1]*

6.2

1 a 92 *[1]*

b 46 *[1]*

2 Centrioles are not essential for mitosis *[1]*; spindle fibres are produced through a different method in plants *[1]*

3 Cancer *[1]*; the normal control of cell division has broken down *[1]*

6.3

1 a Mitosis produces diploid cells from diploid cells/ meiosis II produces haploid cells from haploid cells *[1]*; genetically different daughter cells produced by meiosis II *[1]*

b meiosis I: homologous chromosomes separated during anaphase *[1]*; meiosis II: sister chromatids separated during anaphase *[1]*

2 524 288 *[1]*

3 Crossing over *[1]* (and) independent assortment of chromosomes/chromatids *[1]* create new combinations of alleles *[1]*; (independent assortment of chromatids introduces variation because) crossing over creates genetic differences between sister chromatids *[1]*

6.4

1 Mitochondria for ATP/energy *[1]*; acrosome contains digestive enzymes for entry into egg *[1]*; tail/flagellum for movement *[1]*; protein fibres strengthen tail/flagellum *[1]*

2 Thin/flattened cells *[1]*; present in capillaries (next to alveoli) *[1]* and alveoli *[1]*; reduce diffusion pathway/distance *[1]*

3 Thick cell wall *[1]*; lignin deposits for strength *[1]*; cells are packed closely *[1]*

6.5

1 Both can differentiate into any cell type *[1]*; only totipotent cells have the potential to develop into a whole organism *[1]*

2 They are obtained from adult cells *[1]*; overcomes ethical objections to embryonic stem cell use *[1]*; because iPSCs can be produced from cells of the patient, tissue rejection is overcome *[1]*

3 Mature plants have totipotent cells *[1]* known as meristem cells *[1]*; these cells are able to differentiate into a new plant *[1]*

7.1

1 Relatively high metabolic rate *[1]* requires greater oxygen supply (and carbon dioxide removal) *[1]*; low surface area to volume ratio *[1]* means the diffusion pathway is too long for diffusion alone to be effective *[1]*

2 Good blood supply (i.e. dense capillary networks) *[1]* removes oxygen (and supply carbon dioxide) at a high rate *[1]*; good ventilation *[1]* supplies oxygen (and removes carbon dioxide) at a high rate *[1]*

3 *Actinophrys* (radius = 22.5 μm): SA = 0.00636 μm^2 (6.36×10^{-9} m) *[1]*; V = 4.77×10^{-14} m *[1]*; SA: V = 133,333:1 *[1]*

Actinosphaerium (radius = 200 μm): SA = 0.5024 μm^2 (5.02×10^{-7} m) *[1]*; V = 3.35×10^{-11} m *[1]*; SA: V = 14,997:1 *[1]*

7.2/7.3

1 Short diffusion pathway due to flattened (squamous) epithelia in alveoli *[1]* and close proximity of alveoli and capillaries *[1]*; large surface area to volume ratio *[1]*; steep concentration gradient of oxygen and carbon dioxide *[1]* due to good blood supply (dense capillary network) *[1] (max 3)*

2 12 breaths min^{-1} *[2]* (one mark for correct answer not rounded to two significant figures, e.g. 11.923)

3 **a** Gases in the alveoli and alveolar capillaries/ pulmonary vein are at equilibrium *[1]* because gas exchange occurs until the concentrations on either side of the alveolar walls are equal *[1]*

b The pulmonary artery carries blood to the lungs from tissues (via the heart) *[1]*; oxygen concentration is lower (and carbon dioxide concentration is higher) than in alveolar air because respiration has happened in tissues *[1]*

7.4

1 Exoskeleton limits diffusion across body surfaces *[1]* and some insects have relatively high energy requirements (and therefore high oxygen demand) *[1]*

2 **Level 3 *[5–6 marks]***: Detailed comparisons, with at least three examples of adaptations in each system, explained correctly and linked to function. There is a well-developed line of reasoning which is clear and logically structured. The information presented is relevant and substantiated.

Level 2 *[3–4 marks]*: Includes at least two examples of adaptation per system, explained correctly. There is a line of reasoning presented with some structure. The information presented is in the most part relevant and supported by some evidence

Level 1 *[1–2 marks]*: Simple comments about at least one type of adaptation per system. The information is basic and communicated in an unstructured way. The information is supported by limited evidence.

Relevant scientific points include:

Alveoli: large surface area increases the rate of gas exchange; thin epithelia reduce the diffusion distance; good blood supply maintains concentration gradients; surfactants prevent the collapse of alveoli.

Gills: Large surface area (due to may plated lamellae being present on many filaments) increases the rate of diffusion; thin cell layer in lamellae reduce the diffusion distance; good blood supply in lamellae maintains concentration gradient; countercurrent flow maintains concentration gradient.

3 Countercurrent flow *[1]*; blood in capillaries and water flow in opposite directions *[1]*; oxygen concentration is higher in the water than the capillaries along the length of the gills *[1]*

8.1

1 Diffusion *[1]* because of flatworms' high surface area to volume ratio *[1]*

2 Blood pressure is maintained *[1]*; oxygenated and deoxygenated blood does not mix *[1]*; lower volumes of transport fluid (blood) required *[1]*; blood supply to different tissues can be varied depending on demand *[1]*; delivery of oxygen and nutrients is more efficient *[1] (max 3)*

3 Countercurrent gaseous exchange (see Topic 7.4) in gills *[1]* improves the efficiency of oxygen delivery to respiring tissues *[1]*; body weight is supported in the water *[1]* and fish do not need to maintain their body temperature *[1]*; therefore metabolic demands are relatively low *[1] (max 3)*

8.2

1 Thin walls *[1]*; gaps between cells in the wall *[1]*; high permeability for diffusion across the wall *[1]*

2 Arteries are under higher pressure *[1]*; elastic fibres stretch when pressure is high and recoil as pressure falls *[1]*

3 **a** Veins: 1×10^{-2} m *[1]* arteries: 5×10^{-3} m *[1]*

b Veins have wider lumens to reduce resistance to blood flow at lower pressures *[1]*

8.3

1 Tissue fluid has no large plasma proteins *[1]* and red blood cells *[1]*; tissue fluid has lower concentrations of solutes *[1]*; blood plasma is contained within capillaries *[1] (max 3)*

2 Lymph has less oxygen and nutrients (on average) *[1]* because these particles are taken up by cells prior to fluid draining into the lymphatic system *[1]*; lymph has more leucocytes (particularly T lymphocytes) *[1]*, which are added to the lymphatic system (once they mature in the thymus) *[1]*

3 Capillaries are permeable *[1]* because of thin, single layer of cells/gaps between cells in the wall *[1]*; oncotic/osmotic pressure is always higher in the capillary blood than the tissue fluid (or water potential is higher in the tissue fluid) *[1]*; hydrostatic pressure is high in capillary blood at the arterial end of the capillary *[1]*; hydrostatic pressure decreases towards the venous end of the capillary *[1]*; water diffuses across the capillary wall *[1]* because oncotic pressure outweighs hydrostatic pressure *(max 5)*

8.4

Go further

1 Crocodile haemoglobin contains stronger bonds in its tertiary structure to prevent structural change over a wider range of temperatures

2 Myoglobin is an oxygen store in the muscles. It releases oxygen when respiration rate in muscles is high

Myoglobin's oxygen dissociation curve is to the left of the curve for haemoglobin until approximately 9 kPa O_2 partial pressure

3 Haemocyanin lacks the iron-containing haem prosthetic group, which is responsible for the red colour of haemoglobin. (Haemocyanin instead contains copper)

Summary questions

1 a Haemoglobin's affinity for oxygen decreases as concentration (partial pressure) of CO_2 increases *[1]*

b y-axis = % haemoglobin saturation *[1]*; x-axis = oxygen partial pressure *[1]*; the curve for the higher CO_2 partial pressure is a similar shape but to the right and below the curve for the lower partial pressure

2 Approximately 0.2 *[1]*

3 Sigmoidal/s-shaped *[1]*; haemoglobin gains oxygen in the alveoli (at high partial pressures) *[1]*; oxygen is released at respiring tissues (at low partial pressures) *[1]*

8.5

1 a Relaxation of the heart *[1]*

b Contraction of the atria *[1]*

c Contraction of the ventricles *[1]*

2 The electrical impulse from the SAN (P wave) reaches the AVN *[1]*; the delay is because of the conduction of the impulse down the bundle of His (before ventricular contraction (QRS)) *[1]*

3 Atrial pressure decreases during ventricular systole *[1]*; ventricular pressure is highest during ventricular systole *[1]* and lowest during diastole *[1]*

9.1

1 Central in roots; *[1]*

Towards the outside of stems *[1]*

2 Lignin strengthens the vessels; *[1]*

No end wall, which enables a smooth flow of water; *[1]*

Pits enable water movement between vessels *[1]*

3 Companion cells carry out the functions that the sieve tube elements cannot; *[1]*

Communication between the two types of cell occurs through plasmodesmata *[1]*

9.2

1 Maintaining turgor; *[1]*

For photosynthesis; *[1]*

Transport medium; *[1]*

Cooling plants via evaporation *[1 (max 2)]*

2 Forces water back into cytoplasm/symplast pathway; *[1]*

Because the strip is waterproof; *[1]*

Prevents water returning from xylem to the cortex *[1]*

3 Apoplast: water moves through cell walls; *[1]*

and intercellular spaces; *[1]*

Symplast: water moves through cytoplasm; *[1]*

and plasmodesmata; *[1]*

Water experiences little resistance in the apoplast pathway *[1]*

9.3

1 a Lower rate; *[1]*

Reduced kinetic energy of water molecules *[1]*

b Higher rate; *[1]*

More stomata open *[1]*

c Higher rate; *[1]*

Leaves are more permeable *[1]*

2 Stomata required for gas exchange; *[1]*
But water is lost when stomata are open; *[1]*

3 Radius of tube = 0.5 mm; *[1]*
$\pi r^2 = 0.785$; *[1]*
Volume per minute = 9.42 mm^3; *[1]*
Volume per hour = 565.2 mm^3 *[1]*

9.4

1 *Sources* *(2 max)*
Green leaves; *[1]*
Green stems; *[1]*
Tubers; *[1]*
Tap roots; *[1]*
Food stores in seeds *[1]*
Sinks *(2 max)*
(Growing) roots; *[1]*
Meristems; *[1]*
Developing seeds; *[1]*
Developing fruits; *[1]*
Storage organs *[1]*

2 H^+ ions moved out of companion cells via active transport; *[1]*
H^+ concentration gradient established; *[1]*
Co-transport; *[1]*
Of H^+ and sucrose into companion cells *[1]*

3 Glucose levels are generally lower than other sugars in all tissues; *[1]*
Glucose is converted to sucrose for transport; *[1]*
Relatively high glucose levels around sources in leaves because some glucose is yet to be converted; *[1]*
Sucrose is generally higher than other carbohydrates; *[1]*
Except in tissues surrounding vascular bundles because some of the conversion is yet to occur; *[1]*
Sucrose high in buds, roots, and tubers because it is transported to these sinks; *[1]*
Fructose is low in the leaf blades and sinks because it is not used directly by this plant; *[1]*
Fructose is relatively high in and around vascular tissue, suggesting it can be transported easily *[1 (max 6)]*

9.5

1 Reduced number of stomata; *[1]*
Reduced leaf area; *[1]*
Sunken stomata; *[1]*
Curled leaves; *[1]*
Thick cuticle; *[1]*
Increased water storage; *[1]*
Leaf loss/dormancy; *[1]*
Long roots *[1 (max 3)]*

2 Stomata on the upper surface of leaves (in hydrophytes); *[1]*
The lower surface is often not in contact with the air for gas exchange *[1]*

3 Lab conditions are different from those in the wild; *[1]*
The plant might be damaged when cut; *[1]*
An isolated shoot is smaller than the whole plant *[1]*

10.1

1 **a** The genus name is not capitalised *[1]*
b The binomial name is neither underlined nor italicised *[1]*
c The species name is capitalised *[1]*

2

Taxon	Human	Chimpanzee
Domain	Eukaryota	Eukaryota
Kingdom	Animalia	Animalia
Phylum	Chordata	Chordata
Class	Mammalia	Mammalia
Order	Primates	Primates
Family	Hominidae	Hominidae
Genus	*Homo*	*Pan*
Species	*sapiens*	*troglodytes*

(1 mark per row)

3 Some species reproduce asexually *[1]*; two groups of organisms can be classified as separate species yet produce fertile offspring (e.g. chimpanzees and bonobos), which suggests other criteria can be used to determine species *[1]*

10.2

1 Bacteria have peptidoglycan in their cell walls/fungi have chitin in their cell walls *[1]*; bacteria are always unicellular, whereas some fungi are multicellular *[1]*; bacteria lack organelles *[1]*; fungi can feed using saprotrophic nutrition *[1]*; fungi have hyphae *[1]* *(max 2)*

2 **a** Kingdom = Protoctista, Domain = Eukarya *[1]*;
b Kingdom = Fungi, Domain = Eukarya *[1]*;
c Kingdom = Prokaryotae, Domain = Archaea *[1]*

3 New evidence shows different relationships *[1]*; example of new evidence (e.g. DNA sequencing) *[1]*; two examples of the evidence used in domain classification (e.g. RNA polymerase, cell wall structure, rRNA) *[2]*

10.3

1 New evidence (such as comparative genetics) *[1]*; the evolutionary relationships between species are re-evaluated *[1]*

2 a *Homo erectus [1]*
  b Humans *[1]*

3 Evolutionary positions can be compared/ evolutionary histories can be mapped *[1]*; phylogenetic trees avoid arbitrary groupings of species *[1]*; the use of genetic and evolutionary comparisons enable more precise classification *[1]*

10.4

1 The greater the similarities in the base sequences *[1]* the closer the evolutionary relationship *[1]*

2 Gaps/missing links in the record *[1]*; some species do not fossilise (e.g. due to the lack of a skeleton) *[1]*; some fossils have been destroyed since formation *[1]*

3 Variation exists (between individuals within populations) *[1]*; both artificial selection and natural selection involve individuals with favoured traits breeding *[1]*; in both cases, the favoured traits are inherited by offspring *[1]*; over many generations allele frequencies change within the populations *[1] (3 max)*

10.5

1 a Intraspecific because the variation is within one species *[1]*
  b Interspecific because the variation is between two species *[1]*
  c Intraspecific because similar variation is shown within both species (and no information is provided about how the two species may differ) *[1]*

2 Mutation *[1]* changes DNA base sequence *[1]*; crossing over *[1]* and independent assortment (of homologous chromosomes and chromatids) *[1]* produce new combinations of alleles *[1]*; random fertilisation/mating (increases the number of possible allele combinations) *[1] (4 max)*

3 Identical twins have 100% of the same DNA and fraternal twins have approximately 50% of the same DNA *[1]*; there is a greater genetic influence on height than the risk of strokes *[1]*; the evidence for this is that more than 90% of identical twins have the same height, whereas approximately 15% of identical twins share the risk of having a stroke *[1]*

10.6

1 a Continuous *[1]*
  b Discontinuous *[1]*
  c Continuous *[1]*

2 $(\bar{x}_1 - \bar{x}_2) = 1.6$ *[1]*; $\sqrt{\left(\frac{\sigma_1^2}{n_1}\right) + \left(\frac{\sigma_2^2}{n_2}\right)} = 0.393$ *[1]*; $t = 4.068$ *[1]*; significant difference *[1]*; because p is below 0.05 *[1]*

3 (Note: some calculations for the horn length data are given in the worked example earlier in the Topic). Correctly ranked horn length and number of matings *[1]*; correct rank order differences *[1]*; $6\Sigma d^2 = 90$ *[1]*; $n(n^2 - 1) = 990$ *[1]*, $r_s = 0.909$ *[1]*; the correlation is significant *[1]*

10.7

1 a Physiological *[1]*
  b Behavioural *[1]*
  c Anatomical *[1]*
  d Physiological *[1]*

2 The two species possess similar traits *[1]* as adaptations to similar ecological niches *[1]* but the adaptations evolved independently/at different times *[1]*

3 **Level 3 *[5–6 marks]*:** Detailed description of all three types of adaptation, with relevant examples. There is a well-developed line of reasoning which is clear and logically structured. The information presented is relevant and substantiated.

Level 2 *[3–4 marks]*: Includes at least two types of adaptation. There is a line of reasoning presented with some structure. The information presented is in the most part relevant and supported by some evidence

Level 1 *[1–2 marks]*: Simple comments about at least one type of adaptation. The information is basic and communicated in an unstructured way. The information is supported by limited evidence.

Relevant scientific points include:

Anatomical: flagella for locomotion; pili for conjugation/exchange of genetic material

Behavioural: chemotaxis

Physiological: antibiotic resistance; receptors for binding to cells; relevant enzymes

10.8

Go further

1 The frequency of the lighter fur trait has increased in the population so the allele must have conveyed an advantage. The pale-and dark-coloured mice may still be able to breed and are therefore not separate species.

2 The survival advantage per generation is small, but the cumulative effect over many generations would have resulted in a significant shift in allele frequencies.

Summary questions

1 pH changes *[1]*; temperature changes *[1]*; a lack of water *[1]*; antibiotics *[1]*; destruction by cells in their host's immune system *[1] (max 3)*

2 (For example) nylonase production in *Flavobacterium* *[1]*; this enables industrial waste to be removed *[1]*

3 The insect species has variation between its members *[1]*; the insecticide acts as a selection pressure *[1]*; only individuals that have some resistance to the insecticide will survive, reproduce and pass on their beneficial alleles to their offspring *[1]*; over generations, the proportion of alleles for resistance will increase and the entire population will be resistant to the insecticide *[1]*

11.1

1 *Population*: a group of organisms of the same species; *[1]*

Community: a set of populations within an ecosystem *[1]*

2 Species biodiversity; *[1]*

Because the variety and abundance of species within the ecosystem indicates the stability and health of food chains *[1]*

3 Species richness might be high; *[1]*

But a few species might dominate/many species have low numbers; *[1]*

Biodiversity is based on both species richness and evenness *[1]*

11.2

1 Time; *[1]*

Money; *[1]*

Labour availability; *[1]*

Access (to the ecosystem); *[1]*

Equipment *[1 (max 3)]*

2 Belt transect; *[1]*

Line placed along the ecosystem (from shoreline to the edge of sand dunes); *[1]*

Quadrats used at intervals; *[1]*

Abundance of each species estimated (e.g. by percentage cover) *[1]*

3 Number of samples should be maximised; *[1]*

But will depend on time/resources/number of surveyors; *[1]*

Stratified random sampling; *[1]*

The number of samples in each of the three areas should be proportional to their size (e.g. 100 samples taken in total, 70 on grassland, 20 in the shrubby area, 10 in the orchard); *[1]*

Random sampling is used within each area (e.g. use of a random number generator to select coordinates); *[1]*

Quadrats used to estimate percentage cover of each species *[1]*

11.3

1 To prevent rainwater from entering and drowning the trapped organisms *[1]*

2 Time limitations; *[1]*

In cases where individuals of a species are difficult to count *[1]*

3 22 (6 × 60 × 60, divided by 1000) *[2]*

(one mark for 21.6 not rounded to two significant figures)

11.4

1 *n*: the number of individuals of one species; *[1]*

N: the total number of individuals of all species *[1]*

2 Few species/low numbers in many of the species; *[1]*

An environmental change may cause the loss of species from the ecosystem; *[1]*

Food chains would be disrupted *[1]*

3 Habitat A = 0.500; *[1]*

Habitat B = 0.759; *[1]*

Habitat A has less biodiversity and will be more sensitive to environmental change *[1]*

11.5

1 Small population; *[1]*

Due to disease; *[1]*

Or habitat destruction; *[1]*

Or migration (founder effect); *[1]*

Natural selection *[1 (max 3)]*

2 *Idea that* high genetic diversity indicates a species that has a wide range of traits on which natural selection can act; *[1]*

Idea that when a selection pressure is exerted there is a greater chance that some members of the species will adapt and survive *[1]*

3

Species	Number of monomorphic genes studied	Number of polymorphic genes studied	Percentage of polymorphic genes (%)	Average heterozygosity (%)	
A	28	8	22	4.2	*[1]*
B	35	10	22	7.0	*[1]*
C	8	3	27	5.4	*[1]*

It is not possible to conclude which species has the highest genetic diversity/the data lead to conflicting conclusions; *[1]*

Average heterozygosity suggests species B, % of polymorphic genes suggests species C; *[1]*

However, very few genes have been analysed to determine % of polymorphic genes, especially for species C. Analysing more genes would improve the validity of the results; *[1]*

It is unclear how many genes have been analysed to calculate average heterozygosity. This could be the more accurate of the two measures in this case *[1 (max 5)]*

11.6

1 Desertification; *[1]*

Melting frozen habitats; *[1]*

Changing sea currents/temperature *[1 (max 2)]*

2 (Reduced biodiversity due to) eutrophication; *[1]*

Toxic chemicals poisoning aquatic organisms; *[1]*

Acid rain production altering pH in aquatic ecosystems; *[1]*

Global warming altering sea currents/ temperature; *[1]*

Drainage of rivers and lakes *[1 (max 4)]*

3 Intensive farming increases crop yields; *[1]*

But requires more labour; *[1]*

And more fertilisers/pesticides/herbicides; *[1]*

And exhibits lower animal welfare standards than extensive farming; *[1]*

Intensive farming is more likely to damage ecosystems *[1 (max 3)]*

11.7

1 A greater chance of future drug/genetic resource discoveries; *[1]*

Ecotourism; *[1]*

A greater gene pool for artificial selection *[1 (max 2)]*

2 Reduces the chances of soil degradation; *[1]*

Land remains viable for longer; *[1]*

Biodiversity is maintained; *[1]*

Energy/food requirements can be met for a longer period of time *[1 (max 2)]*

3 Otters prevent an overpopulation of urchins; *[1]*

Which would severely reduce the kelp population *[1]*

11.8

1 Sustainability targets; *[1]*

Targets to reduce desertification/increase land fertility; *[1]*

Targets for greenhouse gas reductions *[1 (max 2)]*

2 Seeds are smaller than plants, therefore easier to transport; *[1]*

And less space is required for storage; *[1]*

Lower probability of disease/damage; *[1]*

Long-term storage/viability maintained for many years *[1 (max 3)]*

3 *In situ* is (usually) cheaper; *[1]*

In situ enables interspecies relationships to be maintained; *[1]*

In situ avoids potential problems when reintroducing species; *[1]*

Ex situ provides optimum conditions/ veterinary care; *[1]*

Ex situ enables controlled breeding programmes *[1]*

12.1

1 Protoctista; *[1]*

Fungi *[1]*

2 **a** Toxin excretion by bacteria *[1]*

b Enzyme secretion by fungi, causing host tissue to be digested *[1]*

3 Viruses can reproduce; *[1]*

But not without exploiting the metabolism of host cells; *[1]*

They cannot synthesise proteins or transform energy; *[1]*

They have evolved over time *[1 (max 3)]*

12.2

Go further

No reliable cure/treatment;

No vaccine;

Further evolution (e.g. if the pathogen evolves airborne transmission) would increase infection rates;

Correct identification of symptoms might be difficult

Summary questions

1 *Similarity*: both involve hyphae penetrating plant tissue; *[1]*

Difference: potato blight is caused by a protoctist, whereas black sigatoka is caused by a fungus; *[1]*

Black sigatoka infects only leaves *[1]*

2 In T helper cells; *[1]*

Reverse transcriptase; *[1]*

Converts viral RNA into DNA; *[1]*

Viral DNA integrated into host cell DNA; *[1]*

Viral proteins and RNA replicated using host machinery *[1 (max 4)]*

3 Genetic mutation resulted in new antigens on the H1N1 virus; *[1]*
New strain not encountered before by human immune systems; *[1]*
No vaccine *[1 (max 2)]*

12.3

1 High population density/overcrowding; *[1]*
Poor nutrition; *[1]*
Poor hygiene/waste disposal; *[1]*
Culture/medical practices; *[1]*
Number of trained health professionals; *[1]*
Lack of public warning systems *[1 (max 4)]*

2 An inanimate object that can harbour and spread pathogens; *[1]*
Examples include:
Meningitis; *[1]*
Influenza; *[1]*
Athlete's foot; *[1]*
Cold sores; *[1]*
Conjunctivitis *[1 (max 2 for examples)]*

3 Soil contamination; *[1]*
With bacteria *[1]*

12.4

1 (Named) antibacterial compound(s); *[1]*
(Named) antifungal compound(s); *[1]*
Anti-oomycetes/glucanases; *[1]*
Cyanide compounds *[1 (max 2)]*
(*Note: insect repellents and insecticides should not be accepted as answers because insects are pests rather than pathogens*)

2 **a** To provide a physical barrier against pathogens *[1]*
b To stop pathogens moving through plasmodesmata to infect neighbouring cells *[1]*
c To stop pathogens moving through phloem sieve tubes to other parts of the plant *[1]*

3 Both are formed from β-glucose monomers; *[1]*
Callose has 1,3 glycosidic bonds/cellulose has 1,4 glycosidic bonds; *[1]*
Both are (largely) linear (although callose has some branches); *[1]*
Callose is helical/cellulose is not helical; *[1]*
Cellulose has cross-links between chains *[1 (max 4)]*

12.5

1 They defend against any pathogen *[1]*

2 *Thrombin*: catalyses the conversion of fibrinogen to fibrin; *[1]*
Thromboplastin: catalyses the conversion of prothrombin to thrombin *[1]*

3 Mast cells release histamines; *[1]*
Body temperature is raised, which inhibits pathogen reproduction; *[1]*
And causes inflammation/swelling *[1]*

12.6

1 Perforin production; *[1]*
Disruption of pathogen cell membranes *[1]*

2 Antibody (complex) with many binding sites for antigen; *[1]*
Pathogens clumped together; *[1]*
Complex is too large to enter cells; *[1]*
Facilitates phagocytosis/phagocytes can engulf several pathogens at once *[1 (max 3)]*

3 T helper lymphocytes infected/destroyed; *[1]*
Cytokines not produced to activate specific B memory cells *[1]*

12.7

1 Antibodies are injected into a person; *[1]*
No memory cells are developed *[1]*

2 Many people in a population have immunity to a pathogen; *[1]*
A disease-carrier is less likely to encounter people who lack immunity *[1]*

3 The tetanus vaccine is in the form of toxoids/modified toxins; *[1]*
Small amounts of toxin can be very harmful; *[1]*
Boosters maintain high levels of antitoxin antibodies and memory cells in the blood (to maintain immunity) *[1]*

▼ *Table of values of* t

Degree of freedom (df)	*p* values			
	0.10	0.05	0.01	0.001
1	6.31	12.71	63.66	636.60
2	2.92	4.30	9.92	31.60
3	2.35	3.18	5.84	12.92
4	2.13	2.78	4.60	8.61
5	2.02	2.57	4.03	6.87
6	1.94	2.45	3.71	5.96
7	1.89	2.36	3.50	5.41
8	1.86	2.31	3.36	5.04
9	1.83	2.26	3.25	4.78
10	1.81	2.23	3.17	4.59
12	1.78	2.18	3.05	4.32
14	1.76	2.15	2.98	4.14
16	1.75	2.12	2.92	4.02
18	1.73	2.10	2.88	3.92
20	1.72	2.09	2.85	3.85
α	1.64	1.96	2.58	3.29

▼ *Critical values for Spearman's rank correlation coefficient,* r_s

	$p = 0.1$	$p = 0.05$	$p = 0.02$	$p = 0.01$
1-Tail Test	5%	$2\frac{1}{2}$%	1%	$\frac{1}{2}$%
2-Tail Test	10%	5%	2%	1%
n				
1	–	–	–	–
2	–	–	–	–
3	–	–	–	–
4	1.0000	–	–	–
5	0.9000	1.0000	1.0000	–
6	0.8286	0.8857	0.9429	1.0000
7	0.7143	0.7857	0.8929	0.9286
8	0.6429	0.7381	0.8333	0.8810
9	0.6000	0.7000	0.7833	0.8333
10	0.5636	0.6485	0.7455	0.7939
11	0.5364	0.6182	0.7091	0.7545
12	0.5035	0.5874	0.6783	0.7273
13	0.4835	0.5604	0.6484	0.7033
14	0.4637	0.5385	0.6264	0.6791
15	0.4464	0.5214	0.6036	0.6536
16	0.4294	0.5029	0.5824	0.6353
17	0.4142	0.4877	0.5662	0.6176
18	0.4014	0.4716	0.5501	0.5996
19	0.3912	0.4596	0.5351	0.5842
20	0.3805	0.4466	0.5218	0.5699
21	0.3701	0.4364	0.5091	0.5558
22	0.3608	0.4252	0.4975	0.5438
23	0.3528	0.4160	0.4862	0.5316
24	0.3443	0.4070	0.4757	0.5209
25	0.3369	0.3977	0.4662	0.5108
26	0.3306	0.3901	0.4571	0.5009
27	0.3242	0.3828	0.4487	0.4915
28	0.3180	0.3755	0.4401	0.4828
29	0.3118	0.3685	0.4325	0.4749
30	0.3063	0.3624	0.4251	0.4670